LAUREN'S PLACE

THE
SCIENCE
OF
STAR WARS

AN ASTROPHYSICIST'S

INDEPENDENT EXAMINATION

OF SPACE TRAVEL, ALIENS, PLANETS,

AND ROBOTS AS PORTRAYED

IN THE *STAR WARS* FILMS

AND BOOKS

THE
SCIENCE
OF
STAR WARS

JEANNE CAVELOS

ST. MARTIN'S GRIFFIN 📖 NEW YORK

To my husband,

who had to pass
the *Star Wars* test
before our first date.

I'm glad you did.

THE SCIENCE OF STAR WARS. Copyright © 1999, 2000 by Jeanne Cavelos. All rights reserved. Printed in the United States of America. No part of this book may be used or reproduced in any manner whatsoever without written permission except in the case of brief quotations embodied in critical articles or reviews. For information, address St. Martin's Press, 175 Fifth Avenue, New York, N.Y. 10010.

Library of Congress Cataloging-in-Publication Data

Cavelos, Jeanne.
   The science of star wars : an astrophysicist's independent
examination of space travel, aliens, planets, and robots as
portrayed in the Star wars films and books / Jeanne Cavelos
     p. cm.
   Includes bibliographical references and index.
   ISBN 978-0-312-26387-4
   ISBN 0-312-26387-2 (pbk)
   1. Space sciences. 2. Star wars (Motion picture) I. Title.
QB500.C38 1999
500.5—dc21   99-22007
                                                      CIP

First St. Martin's Griffin Edition: May 2000

Text design by Stanley S. Drate/Folio Graphics Co. Inc.

D   10 9 8 7 6 5 4 3

CONTENTS

Are planets as common as they seem in *Star Wars*? How likely are Earth-type planets? Could a planet have two suns, as Tatooine does? What conditions are necessary for a planet to develop its own life forms, as we see on Hoth, Tatooine, and many others? What are conditions really like within an asteroid field, and what are the chances of successfully navigating one? How likely is it that humans could live comfortably on so many different planets? How could Naboo have developed its unique structure? Is alien life likely on a moon, such as the Ewok moon of Endor, or the moon of Yavin? Could a planet be entirely desert, like Tatooine? And if so, could life form there?

What might alien life be like? How likely are we to find humanoid aliens? How about intelligent aliens? How likely are the various aliens we meet in *Star Wars*? Why might Jar Jar Binks have those crazy eyes? Why might Wookiees be bipeds? What is life like for a Hutt? Could giant slugs live inside asteroids? Can Ewoks really climb trees? How might Scurriers, Dewbacks, Rontos, and Banthas survive on the desert planet of Tatooine? Might a Sarlacc really take a thousand years to digest its prey? Why do Jawas' eyes glow? What dangers does desert life pose for humans?

Are droids shaped like R2-D2 and C-3PO the most practical types of robots for humans to build? What would be the best design for a battle droid, or for legged tanks like the Walkers? Could we someday build robots as intelligent as Artoo and Threepio? Can we make robots that see, hear, and speak, like *Star Wars* droids? Can we create robots that have their own emotions, and why would we want to? Can robots detect our emotions? Can robots express emotions? Why do Artoo and Threepio constantly argue? Why can't Han and Threepio get along? Can we create bionic limbs like Vader's and Luke's? What injuries might require Vader to use a breathing aid?

ACKNOWLEDGMENTS

I would like to thank my research assistant, Keith Maxwell, for his dedication and hard work. I honestly believe he can find valid scientific research on any topic, no matter how unusual. Keith was an invaluable help throughout the writing of this book.

I would also like to thank the many scientists quoted within this book, who graciously shared their time and expertise and brought their own fascinating perspectives to the *Star Wars* universe.

Special thanks to my Internet group of scientists and *Star Wars* fans. I'm very grateful for the help they provided, brainstorming ideas, serving as sounding boards, contributing expert knowledge, and offering valuable feedback: Tom Thatcher, Dr. Charles Lurio, Dr. Michael Burns, Dr. Stuart Penn, Dr. John Schilling, Dr. Korey Moeller, Elizabeth Bartosz, Dr. Stephanie Ross, Dr. Andrew Michael, Megan Gentry, M. Mitchell Marmel, Dr. Paul Viscuso, Reed Riddle, Carrie Vaughn, Patricia Jackson, John Donigan, Dr. Michael Harper, Janis Cortese, Dr. Michael Blumlein, Joellyn Crowley, Dr. David Loffredo, Beth and Ben Dibble, Jay Denebeim, Bruce Goatly, Dr. Gail Dolbear, Dr. Gary Day, K. Waldo Ricke, Dr. Dennis C. Hwang, Bill Hartman, Patrick Randall, and Margo Cavelos.

Thanks to Sue Gagnon, Mark Purington, and the rest of the staff at Saint Anselm College's Geisel Library, who went to incredible lengths to get me massive amounts of materials in a timely manner, and bravely fought off overdue notices.

Thanks to my editor, Joe Veltre, for his wisdom and enthusiasm, and to my agent, Lori Perkins, for all her support.

Thanks to George Lucas for creating this rich, wonderful universe to explore.

And on the home front, thanks to Igmoe, my iguana, for providing exciting work breaks as he chased me around the house trying to mate with me. And to my husband, thanks for understanding when I only left my computer to eat and sleep (and to run from Igmoe), and for living with all the disorder my work generated. With a little water, my office could look just like the inside of a Death Star garbage masher.

INTRODUCTION

A long time ago in a galaxy far, far away . . .

A high school junior, I walked into a Syracuse movie theater in May 1977. When I walked out, nothing was the same. That opening shot, in which a star destroyer flew endlessly out of the screen, sent my heart racing. I had never before visited such a bizarre, exhilarating, awe-inspiring, fully realized universe. I wanted to live in that "galaxy far, far away." And so I did, for two hours at a time. Many, many times.

I was already fascinated by the idea of space travel, and *Star Wars* fueled my interest in space exploration and the possibility of alien life. As I went through college studying astrophysics, though, I was taught again and again the scientific truths that made *Star Wars* impossible. We cannot travel faster than the speed of light. Life on Earth arose through such an unlikely combination of factors that the chance that all these factors would exist on another planet to create alien life is vanishingly small. Sophisticated robots, when we can build them, will not act human and emotional, but will be logical. And the Force? Pure fantasy.

There was an occasional dissenting voice, but for the most part scientists found the universe of George Lucas incompatible with what they knew.

To be fair, I don't think George Lucas was particularly concerned with science when he created *Star Wars*. Those opening words quoted above sound more like the beginning of a fairy tale than a science fiction saga. And in many ways *Star Wars* feels like fantasy, with the mystical power of the Force; great wizards, called Jedi Knights, who wield it; and great powers of good and evil locked in an epic battle. In creating the part science fiction/part fantasy/part myth that is *Star Wars*, George Lucas did not seek to create a futuristic universe that agreed perfectly with our current understanding of science. If he had, it would have made for some very slow-moving movies. Instead, he sought to combine elements from many different sources and alchemize them into something completely new. And he succeeded.

Yet *Star Wars* contains spaceships, aliens, bizarre planets, and high-tech weapons—all the ingredients of science fiction. These "scientific" elements make the fantastic seem more plausible. Yet how realistic, how possible, is this "galaxy far, far away"?

The answer when *A New Hope* first came out was "not at all." But a strange thing has happened in the years since *Star Wars* first came out. Science is beginning to catch up with George Lucas.

Physicists have come up with theoretical methods of rapid interstellar space travel. Recent discoveries suggest alien planets and alien life are much more common than we previously believed. Many robotocists now think emotions may be a key component in creating intelligent robots. And the Force? A few scientists have theories that can incorporate it.

We have discovered that the universe is a far, far stranger place than we had thought, full of surprises and ever-new mysteries. Say something is impossible today, and you will be explaining how it can be done tomorrow. So while George Lucas may not have attempted to create a scientifically accurate universe, science may actually be turning his vision into truth.

That's not to say *Star Wars* doesn't occasionally screw up big time in its science. Han Solo's boast that he made the Kessel Run in less than twelve parsecs is perhaps the most notorious scientific error in the *Star Wars* films. Since the parsec is a unit of distance, Han is bragging, in essence, that he got to Kessel in twelve miles. Not terribly impressive, if it's twelve miles from here to Kessel.

But the purpose of this book is not to nitpick. The purpose is to scientifically explore the *Star Wars* universe we love so deeply. What can the latest research and theories tell us about making the jump to hyperspace, dueling with light sabers, meeting an alien like Jar Jar Binks, sending an Artoo unit on a secret mission, or levitating an object with the Force? The incredible discoveries science is making can bring us some fascinating insights. And examining the possibilities raised in the movies leads us to some of the most compelling frontiers of science, areas where our very conception of reality breaks down, where mysteries remain unanswered, and where we come up against the ultimate questions of existence.

This book is written so that no particular science background is necessary to understand it. Each topic builds on what has come before, so you'll probably get the most enjoyment out of the book if you read it in order. If you need a reminder about what a particular term means, look in the index to find the page on which the term is first mentioned. That page should provide an explanation. All the measurements in the book are given in the U.S. customary system, with length in feet or miles and temperature in degrees Fahrenheit, unless otherwise stated.

The book covers the four *Star Wars* films: *A New Hope* (originally titled *Star Wars*), *The Empire Strikes Back*, *The Return of the Jedi*, and *The Phantom Menace*. You should be aware, though, that I am writing this book before *The Phantom Menace* is released. I have gathered information about the movie from a variety of sources, but I haven't yet seen it. I also occasionally include issues from the *Star Wars* books or comics, when they seem to shed light on an issue. If you haven't seen the movies recently, you may want to watch them again as you read the book (as if you need a reason to watch them again!).

The book also contains the opinions, expertise, and reactions of some of the top scientists in the fields discussed. What I found in talking with many of them was not only a true love for *Star Wars*, but the belief that a future involving many of the elements we see in the movies may well someday be possible.

When George Lucas first brought *Star Wars* to the screen in 1977, he was a dreamer. After watching *Star Wars*, I—like many fans—wanted to live that dream, to live in "a galaxy far, far away." The discoveries of science may actually one day allow that dream to become a reality.

1

PLANETARY
ENVIRONMENTS

Sir, it's quite possible this asteroid is not entirely stable.

—C-3PO, *The Empire Strikes Back*

I
t comes into view as a small, pale dot against the blackness of
space. Dim, inconsequential beside the brilliance of a star. Yet
for us, it is a safe haven in the endless vacuum of space. Only
here, on this fragile bit of rock or others like it, can life develop
and survive. It formed billions of years ago, the right elements
combining in the right proportions at the right distance from its
sun to bring it to dynamic life. Volcanoes breathed out an atmosphere. Life-giving rains fell, the bit of rock evolved.

As it grows closer, the dot gains color and definition. Major
features are revealed: rock, water, ice, clouds. Within the atmosphere, that protective, nurturing envelope, more details become
apparent. Only on the surface, though, does the unique character
of the planet become clear: the shapes and colors of the topography, the peculiar quality of the star's light scattered through the
atmosphere, the composition and scents of the air, the strength of
the gravity, the texture of the ground beneath our feet, the bizarre
life forms that are another expression of the growth and development of the planet.

We have visited many such balls of life-giving elements. Each
landscape is committed to memory. A flat plain of sand broken

only by harsh, jagged rocks. A vast, snow-covered waste. A fog-shrouded swamp chattering with life. An ancient forest stretching high into the sky. A planet-sized city of level upon level. Some seem mysterious; others feel almost like home. We've seen planets and moons; we've even traveled through an asteroid field. Each has unique characteristics. Anakin's and Luke's home world, Tatooine, is part of a binary star system. Naboo has a bizarre internal structure. The Ewok moon circles the gas giant Endor.

In *Star Wars,* we're swept up in events that take us to a wide array of strange and intriguing planets. They present an exciting picture of the universe as we'd like it to be: filled with exotic yet welcoming worlds. These planets are generally friendly to human life—which is why the human characters have traveled to them. In addition, though, they have indigenous life of their own, in a variety that keeps us surprised and delighted. But how realistic is this view of the universe, based on what we know today? Are Earth-type planets like those we see in *Star Wars* likely to exist? And will so many of them be home to alien life?

YOU CAN'T HAVE AN EMPIRE WITHOUT REAL ESTATE

To have a universe like that in *Star Wars,* the first thing we need is planets, and lots of them. If our solar system is a fluke, and we happen to orbit the only sun in the universe that has planets, then we'll never be able to pop across the galaxy for some Jedi training, set up a hidden base in another solar system, or get into bar fights with intelligent alien life.

How numerous are planets in our universe? Let's first look at how planets form, and what ingredients are necessary in their formation. To form rocky planets like Earth, we need heavy elements like iron, carbon, nitrogen, and oxygen. Unfortunately, they are rare. The two lightest elements, hydrogen and helium, currently comprise 99.8 percent of the atoms in the universe. Hydrogen and helium are great for making stars, but not for creating Earthlike planets or complex life-forms. The heavier elements did not even exist at the beginning of the universe, so stars formed in those early days could not have Earthlike planets orbiting them. Since then, however, stars have been steadily producing heavier elements through the nuclear fusion reactions that power their brilliant light.

In fusion, energy is produced when lighter elements are combined to make heavier ones. When a star exhausts its fuel and dies, it releases these heavy elements into space by exploding or by ejecting its outer layers. A supernova explosion, through its incredible energy, creates even more heavy elements.

If the star lives in a massive enough galaxy, like our Milky Way, then these new heavy elements are held within the galaxy by gravity. They combine with other debris into a cloud of gas and dust, and may eventually form into new stars and planets. These new, younger stars can potentially have Earthlike planets, since the heavy elements necessary have been thoughtfully provided by the older generation.

Considering that *Star Wars* is set "a long time ago," is it *too* long ago to allow for Earthlike planets? While the universe formed about fifteen billion years ago, it wasn't until ten billion years ago that enough heavy elements had been created to form a planet like Earth. Dr. Bruce Jakosky, professor of geology at the Lab for Atmospheric and Space Physics at the University of Colorado at Boulder, concludes that " 'A long time ago' is fine if we're talking a few billion years, but a dozen billion years—that's too long ago." So we've narrowed things down . . . a bit.

Once we have the heavy elements required as raw materials, how do the planets actually form? According to current theory, this debris forms a rotating cloud. Just as a ball of pizza dough, when you toss and spin it, will flatten into a thin crust, so the rotating cloud will collapse into a thin, spinning disk of material. This disk is made up of gas, dust, and frozen chemicals. The dense, inner section of the disk coalesces first into a star. At this point the disk looks like a rotating Frisbee with a hole in the center, the star in the middle of the hole. Dr. Jakosky notes that these disks that form the birthplace of planets seem fairly common. "Between one-quarter and one-half of all stars, when they form, seem to leave behind these disks."

The solid particles in the disk stick together to form large grains of dust. These grains collide with each other and form larger grains, eventually growing into small bodies called planetesimals. A planetesimal may be only a few inches across, or it may be the size of the Moon. Some planetesimals remain small, becoming as-teroids or comets. Others, though, as they rotate around the sun, continue to collide and merge with each other, in a sense sweeping up all the material at the same orbital distance from the sun. As a planetesimal collects all the material in a band around the sun, it becomes a planet. The closer the band is to the star, the smaller the band's circumference is, and so the less material there is to create a planet. That's why, so the theory goes, smaller planets tend to form closer to stars and larger planets farther away.

In addition to affecting planet size, the distance from the star also affects planetary composition. Closer to the star, the disk is very hot, and only materials with high melting temperatures, like iron and rock, are solid. Thus those elements make up the majority of the planetesimals, and the planets. In our own solar system, the four planets closest to the sun—Mercury, Venus, Earth, and Mars—are made up mainly of dense rock and iron. Farther from the sun, where the temperature is lower, additional materials so-lidify, such as water, methane, and ammonia, and become part of the core of the outer planets. These larger planets have stronger gravitational fields, and can attract huge amounts of light gases, such as hydrogen, to surround their cores as massive atmospheres. This process creates distant gas giants like Jupiter and Saturn. Ju-piter, for example, has a core ten times the mass of Earth, which is impressive, but including its thick hydrogen-helium atmosphere, Jupiter's mass totals 318 times Earth's. Each planet, then, is a product of the unique conditions of its formation.

If this theory is true, then planetary formation is a natural part of stellar formation, and there should be a lot of planets out there. Our current theory certainly does a fairly good job of explaining the features we observe in our own solar system. But until re-cently, we've had no other solar systems to test it against.

In the last eight years, however, a string of discoveries has thrown the theory of planetary formation into doubt. Planets seem more common than ever, which supports our theory. Yet the plan-ets we've been discovering around other stars are quite different than those our local system led us to expect. Dr. Jakosky explains, "A lot of the planets we're finding are oddballs." In an attempt to explain the presence of these oddballs, many new theories are being suggested. While most still start with a disk of material orbiting a forming star, many suggest ways in which solar systems much different than our own might result. Why? Because what we're learning is that the universe is a much stranger and more varied place than we imagined.

A PLANET A DAY KEEPS THE EMPIRE AWAY

While science fiction has long posited the existence of other planets, up until recently, we could only guess whether there might be planets orbiting other stars in the universe. False reports of the discovery of planets outside our solar system, called extra-solar planets, have arisen since the 1940s, but only recently have we obtained convincing evidence that such planets do indeed exist.

Planets are very difficult to detect because they're much smaller than stars and they shine only by catching and reflecting a small portion of their star's light. Our sun, for example, is one billion times brighter than the planets that orbit it. If we look at a star through a telescope, the light from the star completely overwhelms that from any planets. As an example of how hard it is to find planets, consider that it took us until 1930 to find Pluto, a planet in our very own solar system. The nearest star, Proxima Centauri, is ten thousand times farther away from us than Pluto. These great distances make seeing planets through telescopes nearly impossible.

Instead of trying to see and photograph extra-solar planets, astronomers instead look for indirect signs of their presence. A wobble in the normally straight path of a star could reveal a star being tugged gravitationally back and forth as a planet orbits it.

We usually think of a planet circling about a stationary star. But the truth is both the planet and the star move, orbiting around their center of gravity. Imagine two children of approximately equal weight—say the twins Luke and Leia at age seven. They face each other, hold each other's hands, and begin to spin around. Since they are of equal mass, their center of gravity will be the point exactly halfway between them, and they will each circle around that point. Their footsteps will trace out a common circle with a common diameter. Now imagine daddy Vader arrives on the scene. He breaks up the circle, turns Luke around to face him, takes Luke's hands in his, and they begin to spin around. Since Vader is much more massive than Luke, the center of gravity will be much closer to Vader. While Vader will not exactly pivot on a single point, he will move off that point by only a small amount, his footsteps tracing out a circle of tiny diameter, while Luke is whipped around in a wide circle.

Just as Vader is not entirely stationary, a star is not completely stationary as a planet orbits it. The planet's gravity affects the star the same way the star's gravity affects the planet. Thus the star will move in a small, cyclical orbit. Our sun's small orbit is generated mainly by Jupiter, its most massive planet. Since Jupiter is one-thousandth the mass of the sun, the sun's orbit is one-thousandth the size of Jupiter's orbit. The sun revolves around a center of gravity just beyond its surface.

Such movements of stars are quite small, so they are very difficult to detect. Yet observing stars has one important advantage over observing planets: stars radiate light that allows us to see them easily. That's why astronomers are searching for planets by looking at stars.

Astronomers have focused on two main techniques for detecting these cyclical movements in a star's course. One is to visually look for tiny wobbles back and forth and measure the extent of these wobbles. This is very difficult, since the wobbles are very small. Let's say the star is the size of our sun and is ten light-years away, and the planet orbiting it is the size of Jupiter. How small would the star's wobble be? Imagine Princess Leia standing two miles away across the flat desert of Tatooine. She plucks a hair out of one of her buns and holds it up. The width of her hair, as it appears from two miles away, is the size of the wobble we're looking for. Not surprisingly, a number of scientists have reported discoveries of planets only to later learn the tiny wobbles they detected were simply observational errors.

A more successful technique has been to search for a cyclical Doppler shift in the light coming from a star. Instead of looking for a wobble back and forth across our field of vision, scientists study the light from a star to see if it is moving toward us and away from us in a cyclical manner. This type of movement causes a shift in the frequency of light coming from the star. Most of us have experienced Doppler shifts—not in light waves, but in sound waves. Imagine a train coming toward you and blowing its whistle in a long, sustained blast. Sound waves will propagate out from the whistle in all directions. Those waves coming toward you, traveling in the same direction as the train, are crunched together by the movement of the train and its whistle. This crunching-up process increases the frequency of the sound waves, making the tone of the whistle sound higher. The train now passes you and starts moving away, still blowing its whistle. The sound waves coming toward you are now traveling in the direction opposite the train, so the sound waves are in essence stretched out. The tone will now sound lower, its frequency decreased.

The same thing happens with light waves emitted by a star. If the star is moving toward us, the frequency of the light increases; if the star is moving away from us, the frequency of the light decreases. Again, these shifts are very small, only one part in ten million if the star has a Jupiter-type planet, and detecting them requires high precision. Yet scientists have reached greater levels of accuracy in measuring Doppler shifts than in measuring visual wobbles. Astronomers can actually measure the velocity of a star toward or away from us down to an accuracy of seven miles per hour. I'm not sure that state troopers are so accurate.

This level of precision means the Doppler technique allows us to find Jupiter-sized gas-giant planets, but not Earth-sized planets, which would cause an even smaller shift. This technique is also much better at finding stars that are moving toward us and away from us at high velocities, when the Doppler shift is greatest. These high velocities are most likely to occur when planets are close to a star. Planets in close orbit revolve around the star faster than planets farther away, forcing the star to also revolve faster. Both the wobble technique and the Doppler technique are most successful at finding large planets around relatively small stars, meaning systems more like Leia and Luke than Vader and Luke, since those solar systems will have the greatest amount of stellar movement.

The first extra-solar planet orbiting a sunlike star was discovered in 1995. Two Swiss astronomers using the Doppler technique found that the star 51 Pegasi moves forward and back every 4.2 days. This means that a planet revolves around the star every 4.2 days: that is the length of a year for anything living on the planet. Since we know that the closer a planet is to a star, the faster it orbits, we know that this planet is very close to 51 Pegasi. In our solar system, the planet closest to the sun, with the shortest orbital period or year, is Mercury. Yet Mercury's year is a leisurely 88 days. The newly discovered planet orbits at only one-eighth the distance from Mercury to the sun. This close, the star would heat the planet to a blistering 1,900 degrees. Not very friendly for life. From our theory of planetary formation, we would expect a planet so close to its star to be small and rocky. Yet the 51 Pegasi planet is a huge gas giant, half the size of Jupiter.

We've now confirmed the discoveries of about fifteen planets around other stars. Most of these are more massive than Jupiter, and most orbit their stars more closely than Mercury. Remember, one reason we've found these "oddball" planets is that they're the easiest to detect. Yet their existence calls into question whether our own solar system is the exception or the rule, and how planets really form. Scientists are struggling to understand how these massive gas giants could have formed so close to their stars, or could have migrated there after their formation.

Not all the planets we've detected are oddballs, though. We have also discovered Jupiter-sized planets farther away from their stars, at an orbital distance comparable to Jupiter, which suggests to some astronomers that Earthlike planets may also exist in these systems. Although we can't yet detect Earth-sized planets around sunlike stars, scientists believe they too may be common. This makes the many Earthlike planets we see in *Star Wars* seem fairly reasonable.

Even with our limited ability to find planets, about one out of every twenty stars we've studied thus far has a planet we can detect. Scientists now estimate that perhaps 10 percent of all stars have planets. That would mean our galaxy alone would be home to twenty billion solar systems. As for how many planets might be Earthlike, we can only make a very rough estimate. But scientists now believe perhaps two billion of these solar systems may have Earthlike planets.

Earth*like*, though, doesn't mean that a planet will look like northern California. It means only that a planet will have a rocky composition and size similar to Earth. Other than that, it may have very little in common with our planet. Mercury, Venus, and Mars are considered Earthlike, but that doesn't mean alien life exists on those planets, or that human life could survive there.

Now that we know planets are plentiful, we need three more ingredients to create the *Star Wars* universe. First, planets that can give rise to their own life. Second, planets that, having the potential to give rise to life, do so. Third, planets that can support human life.

What qualities must planets have to fulfill these needs? Let's consider our first need first.

TWIN SUNS

Luke Skywalker stares off across the Tatooine desert at the dramatic sunset. Two suns, close beside each other, make their way toward the horizon.

Binary star systems, in which two stars orbit around a common center of gravity, are fairly common. Yet scientists think planets around binaries are unlikely, because the gravity of one star may prevent planets from developing around the other. As two stars of different masses orbit about each other, the surrounding gravitational field would shift, setting up potential instabilities in the orbits of any planets.

Even stable orbits would most likely have complex trajectories and variable climates. For example, as a planet orbits past the larger, hotter star, the strong gravitational field would draw the planet close, initiating a period of searing heat. Then as the planet approaches the smaller, cooler star, the weaker gravitational field would allow the planet to swing out to a great distance, sending the planet into a long period of frigid temperatures. In addition, such a planet could have a complex, shifting cycle of sunrise and sunset. This would add to climatic instability.

But astronomers do envision two possible situations in which planets might form in binary star systems, and might even support life. If the two stars are very far from each other—for example billions of miles apart—then planets might be able to orbit one of the stars with minimal influence from the other. For example, Proxima Centauri, the star closest to our sun, is part of a trinary star system. Proxima is one trillion miles from its two sisters, though, 250 times the distance from the Sun to Pluto. Many astronomers believe Proxima could have planets of its own, only minimally affected by its far-off sisters. From the surface of one of these planets, the two sisters would appear only as bright stars in the sky.

The other possibility is that the two stars could be so close together—only a few million miles apart—that to a planet orbiting far enough away, the gravitational field of the two stars would seem almost like that of one. Dr. Jakosky estimates, "If the distance between the stars is only one-tenth the distance to the planet, that would probably be stable." In this situation, the orbit of the planet might be close to circular, and the temperature might remain relatively stable. At dawn two suns would rise, and at dusk two suns would set, just as we see on Tatooine.

Thus, while planets in binary star systems may be rare, Tatooine seems to be an example of one specific situation in which a planet can have a stable orbit. Such planets may well exist, and may even support life. And they'll have some pretty spectacular sunrises too.

ARE STAR SYSTEMS SLIPPING
THROUGH YOUR FINGERS?

"A galaxy far, far away" teems with life. On every planet, over every snowbank, hidden in every cave, submerged in every garbage masher, life abounds. This is one of the qualities that makes *Star Wars* seem so real and so fully imagined. But how common is life in the universe? In a few thousand years, might our descendents be walking into a cantina populated with an incredibly bizarre range of life-forms, a real-life "wretched hive of scum and villainy"?

Scientists believe a wide variety of factors affect a planet's ability to develop life. Many of the necessary characteristics depend not on the planet itself, but on conditions within its solar system, and on the planet's position within that solar system. Here are a few of the key factors.

First, a planet must be at the right distance from its star, where the star heats it to a temperature neither too hot nor too cold, allowing water to exist in liquid form. Planets within a certain narrow range of distances from a star will fall into this favorable band, called the habitable zone. Within this zone, a planet can potentially support life. In our solar system, this zone begins a bit outside the orbit of Venus and continues past Earth to the orbit of Mars. So a planet at Mars's location could potentially harbor life. Yet many factors beside distance from the sun determine the surface temperature of a planet. This means sometimes a planet outside the habitable zone may allow life, and sometimes one within the zone may not. Unfortunately, Mars's small mass—only one-tenth that of Earth—allows it to hold only a thin atmosphere, which keeps Mars cold and prevents liquid water from remaining on its surface. Yet liquid water—and life—may exist below the surface of Mars. We'll discuss this scenario later.

Second, a planet must have time to develop life, particularly complex life. During the prime of its life, the more massive a star is, the faster it burns its hydrogen fuel. Our sun is five billion years old, and if we're lucky it has another three or four billion years in it—time for George Lucas to finish that legendary third trilogy of *Star Wars* movies. High-mass stars, over ten times more massive than our sun, can burn out in only ten to one hundred million years. While some basic life might possibly have time to form, stars with these short life spans most likely won't give their planets time to develop sophisticated life. Dr. Jakosky believes "it's very plausible that life on Earth developed within one hundred million years of conditions allowing it. So you could have life develop in these systems. But there's not time for anything beyond simple single-celled organisms to develop." How long do we need for complex life to evolve? It took four billion years for intelligent life to form on Earth.

While large hot stars are not good candidates for life, stars that are too dim and cool are not good candidates either. Most planets would be too far to receive sufficient heat. And if a planet was close enough to the star to be within its habitable zone, it would face dangerous tidal forces. Tidal forces become a factor in very strong gravitational fields. Since gravitational attraction increases as distances decrease, a planet in close orbit feels a very strong gravitational attraction to the star. In fact, in such an intense field, the gravitational pull on the near side of the planet is significantly stronger than the pull on the far side. If this difference is too great, it will rip any potential planet apart, creating a ring of asteroids. If it's not quite that severe, the difference actually creates a braking force on the planet's rotation. The star literally holds on to the near side of the planet and doesn't want it to rotate away. Thus the planet rotates more and more slowly, its day—the amount of time it takes to rotate once about its axis—growing longer and longer, until ultimately it rotates only enough to keep the same side of the planet facing the sun forever. This means that the planet's day has now become exactly one year long. Although the planet turns, then, for those on the side facing the sun, there is no night. And for those on the side away from the sun, it is always night. Our own moon, in its close orbit around Earth, has succumbed to tidal forces, keeping the same face to us always. A planet that has been braked by tidal forces will be unfriendly to life, with one side extremely hot, and the other extremely cold.

Temperature can be affected by other factors as well. If a planet's orbit is too eccentric, or elliptical, it will undergo dramatic temperature shifts as it moves closer and farther from the sun.

A change in the planet's tilt can also cause climatic changes. French astronomer Jacques Laskar theorizes that the moon helps stabilize the Earth's tilt. The Earth's axis is the line we can draw straight through the center of the planet connecting the north and south poles. The Earth rotates around this line, giving us day and night. At the same time, it orbits around the sun. Yet the Earth's axis does not stand straight up and down, at right angles to the plane of the solar system. Instead, it tilts an average of about 23 degrees. This tilt brings the northern hemisphere closer to the sun in the summer, and takes it farther from the sun in the winter. This allows moderate changes in climate that create the seasons.

The tilt of the axis isn't quite constant, though. It varies slightly, oscillating from 22 degrees to 24.6 degrees and back again every 41,000 years. This small oscillation is believed to be one of the causes of Earth's ice ages. Without the Moon, Dr. Laskar estimates that the Earth's axial tilt would vary from 0 to 85 degrees. This would create huge climatic changes. A tilt greater than 54 degrees would make the poles hotter than the equator. Such an erratic climate would make it much more difficult for life to develop, and for any life to survive long enough to evolve into more sophisticated forms. Planets not lucky enough to have moons might suffer from this instability.

Another danger to the development of life is meteoroid collisions. Meteoroids are chunks of matter in space that enter a planet's atmosphere. Meteoroids can be tiny specks of dust one-ten thousandth of an inch across or huge chunks miles in diameter. They arise from a variety of sources. But most of the meteoroids that fall to Earth are bits of debris from the asteroid belt, a belt of small, rocky bodies that orbits in a zone between Mars and Jupiter. Occasionally these asteroids collide with each other at over 10,000 miles per hour, creating fragments that are propelled toward Earth's orbit. Some meteoroids also arise from the Oort Cloud, a zone of comets that orbits our sun in an extremely eccentric path that takes it as far as two light-years away.

Dr. George Wetherill, planetary scientist at the Carnegie Institution of Washington, theorizes that the strong gravitational fields of Jupiter and Saturn may have scattered trillions of comets and other meteoroids out of our solar system. If that is so, then those planets act as guardians, sparing Earth from many destructive collisions. The presence of bigger planets in a solar system, then, may be necessary for life to develop on smaller planets. Dr. Wetherill estimates that Earth might have suffered one thousand times more collisions if Jupiter and Saturn were not in our solar system.

Although small impacts might not have much effect on us, large impacts can have huge effects on a planet. Scientists believe that sixty-five million years ago, the impact of an object six miles across led to the death of the dinosaurs. Without Jupiter and Saturn protecting us, Dr. Jakosky estimates that Earth might undergo a huge impact like this not every fifty to one hundred million years, but every ten thousand years. "It's hard to imagine how life could survive that extreme an onslaught."

While many of these theories remain speculative, it's clear that a special combination of characteristics are necessary for a solar system to provide an environment friendly to the formation and development of life. We can only assume that the star systems of the life-generating *Star Wars* planets—Tatooine, Hoth, Dagobah, and others—have these characteristics.

Is that likely? Of the two billion Earthlike planets scientists believe may exist in our galaxy, they estimate 2.5 million may be in "friendly" solar systems that offer potential long-term habitability. So the planets we've seen in the movies could easily be in such systems. And if "a galaxy far, far away" contains 2.5 million of these "friendly" systems, we may have a lot more alien species to meet.

Before you start celebrating, imagining 2.5 million planets with alien life, remember that so far we've only discussed the conditions the solar system may need to have for one of its planets to support life. Now we must consider what qualities the planet itself must have.

IS THAT A METEOROID IN THE SKY, OR IS THE EMPIRE JUST TRYING TO FIND ME?

In *The Empire Strikes Back*, the rebels establish a base on the icy world of Hoth. Hoth not only supports the human rebels; it supports its own indigenous life—the Tauntauns, Wampa ice creatures, and others—despite frigid temperatures and heavy meteoroid bombardment. We'll talk about life in icy environments later. For now, let's consider why Hoth might have such heavy meteoroid activity, and what effect this might have on life.

Hoth is constantly bombarded by meteoroids, which fall so often that Luke at first mistakes an Imperial probe for one. We discussed earlier the dangers of meteoroid activity, and the role that Jupiter and Saturn may play in sheltering us from meteoroids.

Even with that shelter, it's estimated that 8,700 tons of meteoritic matter fall to Earth each day. Yet that's spread over such a huge area, and usually falls in such tiny grains of dust, that we're not even aware of it. For the rebels to notice and comment on the high meteoroid activity, Hoth must have significantly greater activity than Earth—where we don't notice it at all—and the meteoroids must on average be larger than those that fall to Earth.

In general, scientists believe the younger the planetary system, the more impacts will occur. If our theory of planetary formation is correct, a young system will have many chunks of rock and ice that haven't yet been captured by planets. The older the system is, the more these bits will have been swept up, either colliding with and being absorbed by planets, or being thrown out of the solar system by gravitational forces.

If we then theorize that the Hoth system is young, however, we run into a couple of big, hairy problems: specifically, the Tauntaun and the Wampa ice creature. These species have evolved on the planet, along with other life-forms we don't see. To develop such complex life-forms, Hoth must be pretty old, probably several billion years. So a young system is unlikely.

If Hoth is a mature planet in a mature system, then why does it have such heavy meteoroid activity? We have two possible explanations.

First, we know that Jupiter and Saturn may shelter Earth from meteoroid impacts. If the Hoth system doesn't have such planets, the rate of impacts could potentially be one thousand times greater than Earth's.

George Lucas provides a second possible explanation in the asteroid belt that orbits Hoth's sun. These are the same asteroids that Han Solo navigates through in a hair-raising attempt to evade Imperial pursuit. As we discussed earlier, the majority of Earth's meteoroids are debris from collisions in our asteroid belt. So perhaps debris from Hoth's asteroid field similarly provides Hoth's meteoroids.

What might Hoth's asteroid belt be like? Luckily, we have a sample in our own solar system, a belt of millions of asteroids that revolves around the sun in a region between the orbits of Mars and Jupiter. Of these millions, fewer than ten are larger than 200 miles across, the size of small moons. A larger number, about 250, are over 60 miles across. Most are $\frac{1}{2}$ mile or less across. While I think the asteroids in the movie tend to be a bit bigger than this distribution suggests, they aren't dramatically different in size from ours.

In our own asteroid field, larger asteroids have compacted themselves into the shape of spheres due to gravitational attraction, while smaller asteroids have irregular shapes. This too is consistent with Hoth's asteroid field. The large asteroid Han lands in is spherical, while smaller ones are in a variety of shapes.

Each asteroid in our own belt rotates, spinning about an axis just as the Earth does. Scientists have measured the rotational period for over four hundred asteroids. They range from 2.3 hours to 48 days, with the average time for a single rotation being 10 hours. Generally, the smaller the asteroid, the longer the rotational period, since these fragments have lost more energy through collisions. The Hoth asteroids are spinning much more quickly, many of them having rotational periods of only 5 to 15 seconds. And the small ones sometimes seem to be rotating faster than the large ones. Since we have not yet observed any asteroid with a rotational period less than 2.3 hours, scientists believe that any asteroid spinning faster than this would be ripped apart by centripetal forces. Perhaps, then, Hoth's asteroids are made of some stronger material.

Even though there are millions of asteroids in our belt, they rarely come within half a million miles of each other. That's because they're spread over such a large area, 6 billion trillion cubic miles. On average, we'll find only one asteroid in a volume of 100 trillion cubic miles. So it wouldn't be much of a challenge to navigate our asteroid field. In fact, several of our spacecraft have passed through it with no harm. The asteroid field Han Solo flies through clearly has many more objects, packed much, much more closely together.

Even in our sparse asteroid belt, collisions do occur. These collisions are infrequent, yet the asteroids have had over four billion years to collide with each other, so many have undergone multiple collisions, as we can tell from the impact craters on them. In the early days of the solar system, asteroids were just some of the many planetesimals orbiting the sun. Some of the asteroids softly collided and stuck together, forming larger objects, just as planetesimals grew to form planets. Yet scientists believe their process of accretion was disrupted by the nearby formation of Jupiter. Its strong gravitational field perturbed the orbits of the asteroids. While most of them had been in similar, circular orbits, many now were drawn into more elliptical orbits, so that rather than bumping gently into each other and sticking together, they now smashed forcefully into each other at 10,000 miles per hour, breaking each other into bits. Thus the asteroids never formed into a planet. The asteroids we currently observe are pieces of larger asteroids that have been broken up, and the process continues, the asteroids grinding each other ever smaller.

We might theorize that some similar gravitational forces are at work in the Hoth asteroid field as well, since those asteroids also failed to coalesce into a planet. Yet the extremely high density of asteroids doesn't seem compatible with the fairly large-sized objects we see. If they were as close as Han's death-defying asteroid run suggests, they would have collided with each other many, many times over, their grinding down process much accelerated from that occurring in our own asteroid field. We have evidence of collisions among Hoth asteroids, in the impact craters on the potato-shaped asteroid and the large circular asteroid Han lands on. If Hoth asteroids collide with velocities similar to those of our asteroids—and some of them seem pretty speedy—then it seems as if they would all be reduced to rubble very quickly. Dr. Charles Lurio, aerospace engineering consultant, estimates, "They'd be sand-sized rubble probably in a lot less than one hundred years." To maintain itself over billions of years, the asteroid belt would have to have much less frequent collisions, like our own, or it would need to balance destructive high-speed collisions with constructive soft collisions. Perhaps conditions in the Hoth asteroid field somehow allow this latter alternative.

If debris from Hoth's asteroid field is raining down with relatively high frequency on Hoth, would all these impacts have an effect on life? Meteoroids 1 mile across or more hit Earth about every 300,000 years. If meteoroids strike Hoth 1,000 times more often, then such an impact would occur every

300 years. A meteoroid this size would explode with the force of the Earth's entire nuclear arsenal. It would leave a crater 10 miles across, and scientists believe it would throw so much dust up into the atmosphere that it would cause major climatic changes, cooling the planet and perhaps even triggering an ice age. In fact, this may very well have happened to Hoth before the rebels arrived. And if it happens again while they're there, well, the Empire won't even have to show up to take care of those pesky rebels.

IS THE UNIVERSE HALF EMPTY OR HALF FULL?

I have good news and bad news. Here's the bad news. Even if a solar system provides a favorable long-term environment for a planet to develop life, the planet must have numerous specific qualities in order to give rise to life. Consider the huge number of favorable conditions that combined to allow the development of life on Earth. This conjunction of so many characteristics must be extremely rare. You might even call the development of life on Earth a cosmic fluke.

What I have just articulated for you is what I call the "pessimists' theory." It dominated scientists' thinking until about ten years ago, and is still subscribed to by many. Let's explore this pessimists' view, in which the development of life on other planets is even less likely than surviving a direct assault on an Imperial star destroyer.

Life arose on Earth due to many factors. For life to arise on other planets, we need these same factors. Here are just a few.

We need a planet with heavy elements such as carbon and oxygen that are the building blocks of organic molecules and so of complex life. Earth is a treasure trove of such elements.

We need a planet with a moderate speed of rotation, which gives us days and nights that are neither too long nor too short. Mercury's day is 88 Earth days long, as is its night. Such long periods of heating and cooling lead to extremes of temperature, 800 degrees during the day, -300 degrees at night. On the other hand, very rapid rotation, which would cause rapid temperature changes, could generate violent weather patterns that would discourage the formation of life.

We need a planet with a strong magnetic field, which protects us from high-energy charged particles in space.

We need a planet with enough heat in its early years—through gravitational contraction, meteoroid impacts, and radioactive decay of elements—to bring it to the melting point. This is how the Earth became separated into different regions, such as the crust, mantle, and core. Currents in our molten core generate the Earth's protective magnetic field and drive volcanic action. Volcanic action melts rock, which contains bound within it huge amounts of water, carbon dioxide, and other chemicals. These are released as gas from the volcanoes, and scientists believe those gases to be the major source of our atmosphere and oceans.

We need a planet massive enough to hold an atmosphere to it, unlike Mercury or the Moon. An atmosphere offers a second line of defense against high-energy charged particles from space and can protect the planet from damaging radiation as well. The atmosphere must not contain a lot of greenhouse gases like carbon dioxide, ozone, or water vapor, which would make the atmosphere trap too much heat. Such greenhouse gases are what make Venus so hot, hotter than Mercury, even though Venus is twice as far from the sun. The temperature is critical, because scientists believe liquid water is necessary for life.

Why is liquid water so critical? Dr. Trent Stephens, professor of anatomy and embryology at Idaho State University, explains, "In order for atoms or molecules to be able to move around sufficient distances and sufficiently rapidly for chemical reactions to occur, we need a medium for them to move around in. Solids don't allow them to move rapidly enough. In gases, the atoms are so far apart, they don't collide sufficiently often to be involved in life. A liquid medium is necessary. For carbon-based chemical reactions to occur, water is by far the best liquid."

Water has a number of unique properties that offer advantages to life. Water is the only material that is more dense in liquid form than in solid, which makes ice float. Such ice can form a protective covering in cold environments, prolonging the existence of water below. Almost half the known elements are soluble in water. This allows water to serve as a medium through which nutrients come to an organism and waste products are removed. Oxygen dissolved in water allows fish to breathe. Nutrients dissolved in water are absorbed by the roots of plants and by the digestive systems of animals. "Water and life go together hand in hand," Dr. Kent Condie, professor of geochemistry at New Mexico Institute of Mining and Technology, says. "You've got to have one for the other."

Some scientists posit that other liquids might serve in place of water, such as methane or ammonia. But Dr. Jakosky explains, "Water is so abundant, since hydrogen and oxygen are abundant, that limiting ourselves to water is not terribly limiting. We expect water to be more abundant on Earth-type planets than any other liquid material."

Not only do we need water, but we need ocean tides. Tidal forces generated by the nearby Moon create ocean tides on Earth, which in turn create tidal pools, and a band of shoreline that is exposed at low tide and covered at high tide. These may have encouraged the formation of life that could survive both in water and on dry land. Dr. Clifford Pickover, biochemist and author of *The Science of Aliens*, believes that without the Moon, "the evolutionary transition to land may have never taken place because the water-land edge would be an insurmountable barrier. The diversity of life on Earth would be reduced fantastically—and humans would not have evolved."

I could go on, and many have, writing entire books on all the factors necessary to the development of life. But you get the idea. In the pessimists' view, we essentially need to reproduce the Earth's environment with only minor deviations in order to generate life. While this may be necessary to produce *Earth*like life, saying it's necessary to produce *any* life is making quite an assumption.

Yet most of the planets we see in *Star Wars do* reproduce the Earth's environment—even more closely than the pessimists would say is necessary. It would seem extremely unlikely that many planets would share all these characteristics. But as we know, scientists have estimated there may be 2.5 million planets with size and composition similar to Earth and a solar system favorable to life. If only a tiny percentage of these come close to reproducing Earth's environment, that could potentially provide the planets we see in the movies.

Pessimists would say that tiny percentage is vanishingly small. In fact, many pessimists believe that all these characteristics came together in the entire history of the universe to create life only once: on Earth.

But the optimists tell a different story. Let's turn, now, to the "optimists' theory," in which life on other planets not only seems likely, it seems inevitable.

A HOME AWAY FROM HOME

Earlier, I mentioned three ingredients necessary for a universe like that depicted in *Star Wars*. We've been talking about the first two: planets that can and do give rise to alien life. In those discussions, the big unknown is how wide a range of conditions alien life can survive. The third ingredient, though, involves a more familiar life form: humans. We need planets that can support human life. But how wide a range of conditions can human life survive?

The movies show us many planets that seem to be settled by humans. For those who want to argue that the people we meet in *Star Wars* are not humans but some alien species with organs and powers unfamiliar to us, I have two points to make. First, we see no sign of any characteristics that would distinguish *Star Wars* people from humans. They look and behave exactly like humans. Second, they are actually called "humans" in the movies. C-3PO's specialty is "human/cyborg relations," and at one point Threepio says of Luke, "He's quite clever, you know . . . for a human being."

In the *Tales of the Jedi* comics, we learn that in the days of the Old Republic, much colonizing of the galaxy occurred. Thus we can imagine that the human species originated on a single world, similar to Earth, and then spread to other worlds. Calling Han Solo Corellian, then, is rather like calling him French. It tells us where he's from, but not what species he belongs to.

When we think of colonizing another planet, most of us generally imagine huge airtight domes within which we would live, sheltered from the unfriendly environment. Yet *Star Wars* shows humans living on many different planets without any isolating domes or breathing devices. How can they survive in these alien environments? There are three options.

First, colonists could potentially be genetically engineered to cope with different environments. If such changes were made, though, then Luke, engineered to live on Tatooine, would have difficulty surviving on Hoth, which has a much different climate. In addition, we see no sign of such genetic differences. If they did exist, then we might think interbreeding among residents of different planets would cause problems, and perhaps even be impossible. Yet Han and Leia don't seem to consider this a concern.

Second, rather than altering the colonists, we might alter the environment. Some scientists believe we will be able to make planets habitable through terraforming, transforming unfriendly planets into Earthlike ones through artificial means. For example, might we be able to give Mars a life-sustaining atmosphere and liquid water? While this would take huge amounts of energy, we can assume that *Star Wars* technology, which allows rapid interstellar travel, has access to such levels of energy. James Oberg, orbital rendezvous specialist for NASA and author of *New Earths*, believes cold Mars, with its minimal atmosphere, might be terraformed through several techniques. Since dark-colored material absorbs more of the sun's energy than light-colored material, he suggests dark soot be spread on the planet's surface. "If permafrost (giant dust-covered glaciers) exists, it might melt, flooding the surface after a billion years of drought." Further heat "could be provided by giant space mirrors, 1,000 km on a side, concentrating sunlight onto the planet." In addition to creating liquid water, the heat could bring frozen carbon dioxide in the polar caps and soil into a gaseous state. The atmosphere might then thicken to the point where plants could be introduced, setting up an ecosystem friendly to man. These ideas remain extremely speculative for now.

In any case, this doesn't seem to be taking place in *Star Wars*, at least not on the planets we've seen so far. For example, if Tatooine were terraformed, substantially changing its temperature and atmosphere, the process would likely kill off any indigenous life forms. Yet we are told that the Jawas, Sand People, and other species are native to Tatooine.

We have one other possibility. Colonists might use artificial aids when initially arriving on the planet but gradually become accustomed to the new surroundings. Such adaptations, which occur during a person's lifetime, are called acquired character. We're extremely limited in how much we can adapt, though. We can put on more layers of fat to protect ourselves from the cold; we can produce more red blood cells to help us acclimate to the thinner atmosphere at high altitudes; we can develop stronger muscles to support us if we gain weight. Similarly, those living on a planet that is colder, has a thinner atmosphere, or heavier gravity could make minor adaptations.

Over generations, colonists might evolve traits that would help them cope in this new environment. This process of evolution, driven by random genetic mutations and survival of the fittest, has occurred throughout Earth's history, as organisms have adapted to new homes or shifting climates. While such evolutionary adaptation may have occurred in limited cases in the very early days of the old Republic, when colonists were isolated on a planet for generations, the galaxy we see in the movies seems very well traveled, with residents of one planet routinely journeying to other planets. Since we see no sign of any acquired characters or evolutionary adaptations in any of the humans in the movies, we must assume that these are minor, meaning these environments are quite similar to the home planet of humans.

Thus we're looking for planets on which humans could survive without any special equipment or adaptations. How likely are we to find a nice selection of those?

Well, they'd need all those qualities listed in the previous section and many, many more just to allow us to survive. Even more than bare survival, though, we'd want a planet to offer a comfortable place to live.

Scientists have estimated that humans wouldn't be comfortable living in temperatures that average higher than 104 degrees or lower than 14 degrees—which may be why Hoth isn't a big vacation spot.

Gravity over 1½ times that of Earth would be tiring to live in. Each action would require more effort, since you would weigh 1½ times what you currently weigh—as if you'd just sat through The Phantom Menace one hundred times in a row, eating three boxes of candy at each showing (actually, that doesn't sound like such a bad idea). Gravity too much lower than Earth's would make you lose muscle and bone mass, leaving you weak and brittle-boned.

The most difficult quality to obtain may be a suitable atmosphere. The idea that another planet's atmosphere might be breathable for us seems unlikely to Dr. Jakosky. "It doesn't take much to make it not breathable. The composition of the Earth's atmosphere is very much an accident of its history. And it's very unlikely we'd find another planet with a similar environment." Dr. Pickover agrees. "This is very unlikely, given how planetary atmospheres are, and how specifically adapted creatures are to the world on which they evolved."

Just as each planet is a product of the unique conditions of its formation and existence, we humans are products of the planet on which we developed, specifically designed to function in this particular environment. Finding another setting in which we are equally at home would be very difficult. These cannot be planets that are simply Earthlike; they must almost exactly reproduce conditions on Earth, which indeed is what most of the Star Wars planets do. As we established in the previous section, a relatively small number of such planets may exist. Yet we're much more likely to find planets that require some minimal use of technology for humans to survive.

All of this leads to another conclusion. Just as man would have trouble living on worlds other than the one for which he was designed, so aliens would have trouble living on worlds other than the one for which they were designed. Thus the Mos Eisley cantina, in which many aliens mix and mingle in a single environment, becomes very unlikely.

But what good is a bar, if your face is enclosed in a breathing mask? Perhaps Darth Vader knows the answer. . . .

"THERE ARE MORE THINGS IN HEAVEN AND EARTH"

While the pessimists' theory was the accepted wisdom for many years, it now seems that all these conditions may not be required for life. What many scientists are coming to believe, based on recent discoveries, is that alien life may be quite different from terrestrial life, and may be able to survive under a much wider range of conditions than previously believed.

In the last twenty years, scientists have been shocked to discover life in extreme environments on Earth that we had previously believed inimical to life. We've discovered primitive organisms living over two miles underground on only water and rock; thriving in boiling water from super-hot volcanic vents on the ocean's floor; and luxuriating in the frigid Arctic Ocean. These bacterialike organisms are called extremophiles because they thrive in extreme environments. Scientists now believe that thousands of undiscovered extremophiles exist on Earth.

These organisms are a lot more than curious oddities. In fact, they may have been among the first life-forms to develop on Earth. You've probably heard the theory that life on Earth originated in a "chemical soup" with lightning triggering chemical reactions. Yet many scientists these days believe life on Earth originated around underwater volcanic vents, called hydrothermal vents. These vents shoot out superheated water and minerals, providing energy that helps trigger the formation of organic molecules and encourages the growth of more complex life. Since Earth underwent heavy meteoroid bombardment during the early part of its existence, the Earth would have been much hotter then than it is now, with conditions like those around hydrothermal vents more common. Many scientists believe that heat-loving extremophiles may have been the first life on Earth four billion years ago and the ancestors of all life.

The discovery of life in such extreme environments suggests that life may also exist in similar extreme conditions on other planets. While the surface of Mars seems a dry, frigid wasteland, Dr. David McKay, senior scientist for planetary exploration at NASA's Johnson Space Center, believes that conditions about a mile below the surface of Mars may be quite similar to those deep underground on Earth, with geothermal heat from volcanic activity and a groundwater system. "Underground on Mars is a good place to look for life."

Possible evidence that life existed on Mars several hundred million years ago was uncovered in 1996. Dr. McKay and colleagues claimed to have discovered organic molecules and possible fossilized bacteria in a Martian meteorite found in Antarctica. While the evidence is still controversial, the rock contains a combination of elements that strongly suggest life. "I am convinced," Dr. Stephens says, "at least tentatively, that the meteorite from Mars really does have evidence of life in it."

If life can survive and thrive in a wider range of conditions than we thought, then we no longer need other planets to reproduce Earthlike conditions in order to give rise to life. In that case, we can reduce the requirements for the development of life to three: complex organic molecules, water, and energy. To be considered organic, a molecule need not be created in a biological process; it simply needs to contain carbon, hydrogen, and oxygen. Dr. Jakosky believes we needn't even require the existence of organic molecules, but only of the elements out of which these could be built.

With this new outlook, we see preliminary indications of life everywhere. We've discovered complex organic molecules on the outer planets of our solar system, on comets and meteorites, in interstellar dust, and even in other galaxies. "There's carbon galore out there," Dr. Stephens says. Some scientists believe such molecules, carried by extraterrestrial sources, introduced the building blocks of life to Earth. Ten percent of the meteoritic material landing on Earth is made up of organic molecules, "seeding" the planet with the basic ingredients for life. Some meteoroids even carry amino acids, the building blocks of proteins. And Earth isn't the only planet receiving these "seeds."

Many scientists believe ice-covered Europa, a moon of Jupiter slightly smaller than our Moon, may also harbor life. While Europa has a very thin atmosphere that includes oxygen, the atmosphere doesn't seem sufficient to support life on the frigid surface. Since Europa is far outside our sun's habitable zone, we used to believe that the planet was covered with a 150-mile-thick layer of ice. Yet data from the Galileo probe in 1996 revealed that liquid water may exist beneath a relatively thin layer of ice, forming an ocean that may be a breeding ground for life. Since Europa gets very little light and heat from the sun, how could liquid water exist? Heat from radioactive elements in Europa's core and tidal gravitational forces caused by Jupiter and Jupiter's other moons most likely drive geological activity and heat the planet, preventing the water from freezing. Scientists now theorize that Europa's internal heat may have created hydrothermal vents on the ocean floor just as it has on ours, and that these vents may have stimulated the formation of life, just as we believe they did on Earth. Dr. John Delaney, an oceanographer at the University of Washington, feels sure that life does exist on Europa. If life can exist on Europa, imagining life on Hoth is easy.

Alien life may currently exist, then, on two other bodies in our solar system. Scientists also believe life may have existed in the past on additional bodies, particularly Venus and Saturn's largest moon, Titan. Since we believe that life either currently exists or may once have existed on four different planets and moons in our solar system, life in other solar systems seems fairly likely.

Another piece of evidence in support of the optimists' theory is that scientists now believe life on Earth formed much earlier than previously believed, soon after conditions arose that made the existence of life possible. Around 4 billion years ago, near the end of the Earth's formation process, the bombardment of Earth by planetesimals finally slowed. Evidence now indicates that life may have existed as long as 3.85 billion years ago, and that life was widespread by 3.5 billion years ago. This means life took only a few hundred million years to form. Dr. Jakosky believes that this rapid development "suggests that the origin of life must be a very straightforward, natural process. Any planet that has the environmental conditions we think are necessary for life should undergo an origin of life. It's hard to believe it wouldn't be common any place you have the ingredients. The most likely outcome is that life is widespread in the universe."

The optimists even have an answer to the pessimists' view that our solar system, and our Earth, embody a unique combination of characteristics that is extremely unlikely to arise again. Why, the optimists ask, is our particular solar system so unique? We orbit a fairly average star in a rather unremarkable section of the galaxy. So rather than saying that our solar system is somehow unique, isn't it more reasonable to say that similar stars in similar neighborhoods might similarly develop life? If life originated once, the optimists argue, then it probably originated multiple times. And their view is now becoming the dominant one among scientists.

Dr. Jakosky believes that "life is common," and feels we're on the verge of a great breakthrough. "It's been just within the last few years that we've discovered planets and we've discovered the potential for life to exist on other planets. That makes this a special time in the history of humanity. We're the first generation that has a realistic chance of discovering life on another planet." Dr. Stephens is also hopeful about the existence of alien life. "Some biologists think we're the result of a series of accidents. I think that idea is absolutely preposterous. It has led us to believe our universe is sterile except for this tiny insignificant corner of the universe where we just happened to be lucky. I don't think that's the case at all." This new vision, of a universe filled with life, seems to reflect perfectly the *Star Wars* universe, in which life has developed in every available niche.

JOURNEY TO THE CENTER OF NABOO

Naboo is a puzzle. According to the official *Star Wars* website, the planet is made up of large rocks thousands of miles in diameter. These rocks don't fit together terribly neatly, creating a honeycomb structure of caves throughout the planet's interior. This extremely unusual structure defies natural explanation. The most likely explanation, actually, is that the planet was artificially created, either built from scratch like the Death Star or altered through some sort of massive mining operations. But the description doesn't indicate any artificial components or alterations. Let's consider, then, the factors that would be involved in the natural formation of such a planet.

The major difference between Naboo and other planets seems to be that the many rocks that comprise it never melted and fused into a unified planet. Naboo remains a conglomeration of separate rocks, held together by gravity. Since planets are formed by an aggregation of planetesimals, it makes sense that Naboo is made up of many rocks. Yet when planetesimals collide, like asteroids, they will either collide at high velocities and break each other apart, or collide at low velocities and stick together. When they stick together, much of the kinetic energy of their motion is converted into heat energy. This heat fuses the bodies together. If the planetesimals forming Naboo failed to fuse, the heat generated must have been minimal, which means the planetesimals must have been traveling at nearly the same velocity in the same direction when they gently touched together.

While two small planetesimals might possibly come together like this, we get in trouble when the proto-Naboo has grown larger. As the body that will eventually become Naboo grows larger, its gravitational attraction will grow stronger. It will draw planetesimals to it with increased force, and they will accelerate toward it and impact it with higher velocity. Heat is an unavoidable consequence.

Thus the only way these rocks can come together without generating significant heat is if they all come together simultaneously, at very low relative velocities. This would require some bizarre gravitational forces to be herding the planetesimals together, forces that are hard to imagine.

If the planet does somehow form in this way, would it remain this way? Most planets differentiate into a core and separate layers. As we know, the early Earth was heated by gravitational contraction, meteoroid impacts, and radioactive decay, bringing it to the melting point. This is how the Earth separated into a core, mantle, and crust.

Some of these heat-generating elements may have been scarcer on Naboo than Earth. Naboo may have suffered fewer meteoroid impacts than Earth, and it may have a smaller fraction of radioactive elements in its rock. This could possibly explain why it never reached the melting point and differentiated.

Yet Naboo doesn't seem to have particularly light gravity, so we can assume it has a mass similar to Earth's. A planet with the mass of Earth will feel a strong gravitational attraction inward. Every rock in that planet will be drawn toward the center, which is why the planet stays together. And the deeper a rock is beneath the surface, the greater the downward pressure it feels from all the material above it, all pushing toward the center.

This pressure of the outer layers on the inner ones is what causes large asteroids and planets to contract into a spherical shape. If a cave within a planet is made of materials that can't withstand this pressure, it will collapse. In the case of Naboo, it seems likely that these caves would have collapsed, particularly those a significant depth beneath the surface. The Naboo rocks must have some special structural strength that allows them to withstand the pressure, and that has kept them from contracting and melding with one another.

The special structure of Naboo also makes it difficult to understand how it can have liquid water or an atmosphere. Scientists believe that both of these arose on Earth through volcanic processes. Without enough heat to melt the rock and close those gaps, it seems unlikely that Naboo would have volcanism.

An alternate theory posits that a planet's water might come from comets impacting the surface. Comets, often described as "dirty snowballs," carry a fair amount of ice. Yet recent scientific observations have thrown this theory into doubt.

Whatever the answer, Naboo must certainly have arisen from a strange combination of factors. And how a planet so different from Earth might have developed an environment so similar remains a mystery.

MOON OVER ENDOR

Now that we know some of the conditions necessary for life, let's see if those conditions are satisfied in a couple of the places where we find life in *Star Wars*.

Around the silvery gas giant Endor, nine moons orbit. The largest of these, the size of a small planet, is home to an ancient forest and a species of diminutive furry creatures called Ewoks. But is a moon a likely home for the Ewoks? Are moons likely homes for life?

If our own solar system is any indication, moons are fairly common. Except for the two planets closest to our sun, Mercury and Venus, all the planets in our solar system have moons. We have a total of over sixty. Saturn has the most with eighteen. The majority of our solar system's moons are quite small, though—as small as ten miles across. These may be asteroids that were scattered out of our asteroid belt and gravitationally captured by the planets. Such tiny moons are too small to retain an atmosphere or give rise to sophisticated life. Larger moons are less common. Saturn has only one, Titan. Jupiter has four: Io, Europa, Ganymede, and Callisto. The largest moons—Titan, Ganymede, and Callisto—are actually about equal in size to Mercury. It seems reasonable, then, that the gas giant Endor has nine moons, and that one of them may be of significant size. In our solar system, Titan is the only moon with a thick atmosphere, so a life-giving atmosphere, while not common, is at least possible.

What would conditions be like on such a moon? If the Ewok moon was made of the same materials as Earth yet was the size of Ganymede, it would have gravity only 7 percent that on Earth. This would take some getting used to, since if you weighed 150 pounds on Earth, you'd weigh only 10.5 pounds on Endor's moon.

Since the gas giant Endor is quite massive, the moon would likely be subject to tidal forces. Presumably the Ewok moon is not so close to Endor that tidal forces would rip it apart. Yet even so, the forces would slow the rotation of the moon. As we discussed earlier, tidal forces have braked the rotation of our Moon so that it keeps the same side facing the Earth at all times. Ganymede similarly presents the same face always to Jupiter. If this is also true of Endor, those who live on one side of the moon would always see Endor in the sky, undergoing changes over the course of a day like the phases of the Moon. Those who live on the other side of the moon would never see Endor. Being in this sort of tidal lock is much less serious for a moon than it is for a planet. If a planet is tidally locked about the sun, one side will bake while the other freezes. In this case, although one side of the moon always faces Endor, both sides will receive sunlight. The cycle of light and dark will not depend on the rotation of the moon about its axis, but instead on its revolution about Endor.

One day on the Ewok moon would be the time it takes to orbit once around Endor. The large moons of Jupiter take from 1.7 to 16.7 Earth days to orbit their planet. The farther the moon is from the planet, the longer its period of revolution. As we discussed, the best chance for complex life on the Ewok moon occurs if it has a day only a few Earth days long. A longer day would most likely result in serious temperature extremes and violent weather patterns unfriendly to life. So the Ewoks would be better off with a moon closer to Endor.

Tidal forces could also provide energy potentially helpful to life. They would generate internal heat within the moon, as we believe is happening on Europa. This internal heat could help generate volcanic action. Volcanoes could then vent water vapor into the atmosphere and ultimately lead to the creation of oceans, as we believe happened on Earth.

But would the temperature on the planet allow liquid water to form? For an Earthlike environment like the one we see on the Ewok moon, Endor would need to orbit within the habitable zone of its star, as Earth does. This would put any moons into the habitable zone as well. Endor is a gas giant like Jupiter and Saturn, though, and they orbit far beyond the habitable zone. But as we know, in the past few years scientists have discovered gas giants orbiting other stars much more closely than we believed possible. This suggests that a gas giant within the habitable zone is possible. Thus life on such a moon seems quite believable. And if gas giants in close orbits are as numerous as we now believe, life on moons around those gas giants may be more plentiful than life on Earthlike planets.

Life on the Ewok moon may not be all cuddling and singing, though. One problem the Ewoks would face is radiation. Jupiter has lethal radiation belts like Earth's Van Allen belts, only Jupiter's are much larger and more intense. These radiation belts are generated by Jupiter's powerful magnetic field, which is about ten times stronger than Earth's. The magnetic field traps high-energy electrons and protons in space, causing them to spiral back and forth at high velocity along the magnetic field lines.

To picture the magnetic field lines, you might recall an experiment in science class, in which you scatter iron filings onto a sheet of paper and then put a bar magnet onto the paper. The filings move to arrange themselves along the magnetic field lines, which emanate upward out of the north pole of the magnet, bow out around the side of the magnet in a semicircle, and come together again at the south pole of the magnet. Now imagine that bar magnet standing upright within the center of Jupiter. The magnetic field lines emanate from the north pole, bow out into the space surrounding Jupiter, and then come together again at Jupiter's south pole. Now imagine charged particles spiraling rapidly around these lines.

While the magnetic field keeps most of these particles away from the planet itself, it's bad news for anything nearby. When the Galileo space probe traveled into one of Jupiter's radiation belts, its heat shield was penetrated by a million high-energy particles per second.

The four inner moons of Jupiter travel through Jupiter's radiation belts and are bombarded with these particles. Such high-energy particles can rip through matter, breaking apart molecules—including DNA—and atoms. Such radiation is extremely harmful to humans and almost all living creatures. Dr. Jakosky says, "A terrestrial life form that was transplanted to the surface of Europa would not be able to survive for very long at all due to the breaking of the chemical bonds." Yet there is an exception to this rule. Some hardy terrestrial bacteria can withstand extremely high levels of radiation. In fact, Dr. Jakosky points out, "They live in the water inside of nuclear reactors!" Although they sustain serious genetic damage, they have special methods for coping with this damage and rapidly repairing it. If the Ewoks have similar abilities, they might be able to survive in such an environment. I wouldn't hold out much hope for Luke, Vader, and all the non-natives, though!

Another possibility—one more friendly to tourists—is that the atmosphere of the Ewok moon shields the surface from radiation. Our own atmosphere provides a similar shield. The atmosphere of the Ewok moon would have to protect against an even greater radiation threat, but it could potentially keep its furry little occupants safe.

One further possibility. If a ring of rocky material orbits between Endor and the Ewok moon, it may absorb the charged particles, clearing the area beyond. A ring around Saturn has been shown to reduce the number of charged particles. And a ring of space debris around Earth is being found to create a similar effect.

The key factor in all of this is the distance of the moon from

Endor. The moon would need to be close enough to have a short enough day and internal tidal heating; yet it would need to be far enough to prevent tidal forces from ripping it apart and to keep radiation to a minimum level. The distance from Endor to the moon can't be too far or too close; as all bears know, it has to be *just right*.

THE "BRIGHT CENTER TO THE UNIVERSE"?

The planet we've had the most time to visit is Tatooine. It feels so familiar now, I almost feel like I've been there.

Home of Anakin and Luke Skywalker, Tatooine is a harsh planet and appears to have no open bodies of water. Violent sandstorms sweep a landscape of desolate sand dunes, sharp mountains, narrow canyons, and barren rocky wastes. Water apparently exists beneath the surface of the planet and as water vapor in the air.

While we don't know all the details of Tatooine's environment, we do have similar environments on Earth: deserts. The defining characteristic of a desert is the small amount of precipitation it receives: less than ten inches per year. The area of southern Tunisia where parts of *Star Wars* were filmed receives on average only six inches of rain per year. Tropical rain forests average 80 inches per year. Arabs call the desert the "sea without water." This is echoed in the name of Tatooine's "Dune Sea," the undulating field of sand dunes where *Return of the Jedi*'s Sarlacc lives.

This lack of water provides the main challenge to life. In some parts of the Sahara, no rain has fallen in more than twenty years. Then suddenly rain falls in a torrent, flooding across the landscape, plowing boulders ahead of it and cutting channels into the ground before it finally sinks into the subsoil. We see signs of similar rain erosion on Tatooine, which has several dry riverbeds: one where Artoo is captured by the Jawas, and another by Jabba's palace.

In the desert, the land is so dry that the air is often parched, with only 2 to 5 percent humidity. Clouds, then, are rare, and sunlight beats directly down, with up to 95 percent of it reaching the ground. In midday, heat-absorbing black rock can reach 185 degrees, way too hot to walk on. Since water serves as a moderating influence on temperature, areas without water heat and cool much more quickly. At night, the temperature can plummet to only 50 degrees. Such extreme temperature variations actually cause rock to crack and burst. As Luke Skywalker stands in the heat of the day on a mountain cliff above Mos Eisley, he may hear a series of loud cracks like cannon fire as the mountain rock fractures. As rocks break off and fall down the mountains, they shatter into smaller pieces. Weathering reduces these fragments to sand and even smaller dust.

Another phenomenon Anakin and Luke may be familiar with is ghost rain. In the desert, a hot layer of air often forms just above the ground. Rain may begin to fall, but when it hits that heated layer, it evaporates, never reaching the ground.

Contrary to the vision you may have of it, a desert is not all sand. Sand and dust are simply the endpoints of erosion of rock by heat and wind. The Sahara is only 20 percent sand. The remainder includes mountains, plateaus, piles of slag, gravel-covered wastes, ravines, and canyons. We see many of these features on Tatooine. The Jawas hide in caves in a rocky cliff face. Luke catches up with the renegade Artoo in a canyon, and Threepio falls off a plateau when the Sand People attack.

In *The Phantom Menace*, more of Tatooine's topography is revealed in a white-knuckle pod race. The race course seems to be part of a network of deep twisting ravines and crevasses that includes some bizarre stone formations, such as a series of natural stone arches. The Sahara's Tassili n'Ajjer Plateau, an area of southern Algeria, provides a striking parallel to this part of Tatooine. The Tassili Plateau is made up of hundreds of huge blocks of sandstone that split apart long ago. Narrow steep-sided ravines, up to 2,000 feet deep, separate these blocks. Numerous shallower canyons, up to 200 feet deep, crisscross the plateau. German naturalist Uwe George, author of *In the Deserts of This Earth*, says, "The canyons run through the plateau like the streets of a large city, with 'apartment houses' between them." Within this city are dead ends, intersections, tunnels, caves, stone needles 100 feet high, and bridges arcing like overpasses over the streets. These bizarre formations were caused by wind and water erosion, the water erosion taking place long ago when the Sahara had a wetter climate, long replaced by wind erosion in this extremely arid region. Since the wind carries sand and dust, high winds, according to Uwe George, act like "a sandblasting machine." It's certainly reasonable to find such formations on Tatooine. They also suggest that there may once have been more water on Tatooine than there is now.

Before we try to figure out what Tatooine may have been like in the past, let's examine its current condition. As far as we can tell, the entire planet appears to be a desert. Earth has a much more varied climate, with only 30 percent of the continents being desert. Deserts form in particular areas because of a variety of factors. Most terrestrial deserts lie near the Tropic of Cancer or Tropic of Capricorn, where the sun's rays are at their strongest during the Northern Hemisphere's summer and the Southern Hemisphere's summer, respectively. Patterns of air circulation, ocean currents, and elevation also contribute to the formation of deserts. Once formed, deserts tend to expand. Strong winds dry out the topsoil of neighboring lands and blow it away. Sandstorms cause even more dramatic erosion. They can sand the paint off a car and wear the landscape down to bare rock. The deserts on Earth are expanding in this way, with the Sahara growing by forty square miles each day.

Even if Earth's continents did become huge deserts, though, our planet would still have one major difference from Tatooine: oceans. We don't see any large bodies of water on Tatooine, and unless they're hiding on the far side of the planet, we have to assume there are none. Dr. Condie finds the lack of surface water hard to accept. "To have a whole planet that is a desert and yet to have moisture in the atmosphere I think would be difficult."

As we discussed earlier, scientists believe most of the free water on Earth came from water locked in the rocks, and since Tatooine generally seems quite similar in composition to Earth, we can assume that Tatooine similarly has water locked in its rock. On Earth, that water was liberated by volcanic action. On Tatooine, then, less volcanic action may have freed smaller quantities of water.

Why would Tatooine have had less volcanic activity? Perhaps Tatooine has smaller quantities of radioactive elements that serve to heat it. Or if the composition of Tatooine is about the same as

Earth, then perhaps Tatooine is less massive than Earth. A less massive planet would have less internal heat. An example of such a planet is Mars, which is only about half the size of Earth. Mars has some volcanic activity, but it has been decreasing steadily, and Mars has never been as geologically active as Earth. Thus it's not too difficult to explain why a planet would have less free water than Earth. With this theory, Tatooine always would have had less free water, and it always will have less free water. Signs of water erosion, then, would simply be the result of irregular downpours, as occur in the desert.

In proposing ways Tatooine differs from Earth, though, we have to be very careful. As discussed earlier, terrestrial life survives on Earth because of many different factors in delicate balance. A planet that differs significantly from Earth in any way will likely not be able to support human life. For example, if Tatooine had the geologic activity of Mars, human life could not survive on it. Remember that currents in the liquid outer core create a planet's magnetic field. Because of its minimal internal activity, Mars has a very weak magnetic field, which does not keep damaging high-energy charged particles away from the surface as Earth's magnetic field does. So we must propose only incremental differences between Tatooine and Earth, slight changes that might be just enough to create the differences we see. Tatooine, after all, is much more like Earth than any of the planets in our solar system. While we're doing this, please be aware that any theories we consider will be very speculative. A planet is an extremely complex system, with many different factors interacting, and we don't yet fully understand how they all affect each other. After all, when it comes to planets, we only have a few examples to work with, and only one that we've studied in detail.

If we want to theorize that Tatooine had more water earlier in its history—and accounting for signs of erosion is only one reason we might want to do this—we need to work a little harder. One of our local planets provides a handy comparison. While Mars is now a desert planet, it is covered with signs of water erosion. Scientists believe Mars once may have been warmer and wetter. But what happened to the water? Current theories involve several different processes occurring together. One of these in particular might be relevant to Tatooine.

Scientists believe that some of the water on Mars has simply evaporated from its atmosphere into space. A planet holds an atmosphere to it by gravity. The molecules in the atmosphere are constantly moving around, colliding with each other. The higher the temperature, the faster the molecules move and the more they collide with each other. In these collisions, a lighter molecule can pick up more speed than a heavier one. A lighter molecule may then actually attain the escape velocity required to overcome the planet's gravitational attraction. It will escape the atmosphere and head off into space.

Molecules are constantly leaking from the atmosphere of any planet into space. The less massive the planet, the lower the required escape velocity. Dr. John Schilling, research engineer at SPARTA, Inc., explains, "The occasional molecule in the upper atmosphere picks up enough speed in a 'lucky' series of collisions to escape into interplanetary space." The rate of evaporation depends on the mass of the particular molecule, the mass of the planet, and the air temperature. These factors allow light elements like hydrogen and helium to escape from Earth's atmosphere. In the weaker gravity of Mars, molecules as massive as water can sometimes escape, even though the temperature, and so the velocity of the molecules, is lower.

We really don't want Tatooine to have gravity as weak as that of Mars. Mars, after all, has only a very thin atmosphere and can't support complex life. If water molecules can escape the atmosphere too easily, we won't have any water left, and we'll lose other elements as well.

Yet we don't need to lose an entire water molecule to lose water. Ultraviolet radiation from the twin suns can strike a water molecule in the upper atmosphere and break it into hydrogen and oxygen. The light hydrogen could escape, while the heavier oxygen remains trapped in the atmosphere. The water has virtually been lost, though the entire molecule did not leave the planet. This process occurs on Earth, though at a very low rate, since our atmosphere acts as a shield against most ultraviolet radiation. Only if a molecule makes it to the upper part of the atmosphere will it likely be hit by ultraviolet rays.

If we theorize that Tatooine is a bit less massive than Earth, then it might start out with less free water in the first place. Water molecules in the atmosphere could rise higher in the lighter gravity, more easily reaching the upper portion where they would be exposed to ultraviolet radiation. We could even imagine that Tatooine's twin suns put out a bit more ultraviolet radiation than our sun—though not too much, unless you want Luke and Anakin to get skin cancer. Another factor could contribute as well. If Tatooine does have a magnetic field just a bit weaker than Earth's, it would offer slightly less protection from high-energy charged particles in space. These particles could also break apart water molecules. Thus we could explain a very slow loss of water from the atmosphere, leading to a gradual drying of the planet over billions of years. Humans happened to find and colonize Tatooine during the few tens of millions of years when it had lost most of its moisture but still had enough remaining to make it habitable.

So if we're theorizing that Tatooine is a "bit less massive" than Earth, how massive is that? Mars, at about half the size of Earth and just 10 percent the mass, seems too small. Dr. Michael Burns, a theoretical astrophysicist and president of Science, Math, and Engineering, Inc., believes a planet about two-thirds the size of Earth, with a mass about 30 percent that of Earth, might fit the bill.

Even though Tatooine may be very slowly losing water, at the time we see it, it still retains a fair amount. Moisture exists in the air of Tatooine. We see clouds in the atmosphere, and Luke lives on Owen and Beru's moisture farm, which, according to the *Star Wars Encyclopedia*, condenses moisture from the air. What water Tatooine has, then, would circulate through Tatooine as it does in the desert. Water vapor in the atmosphere would give rise to ghost rains and rare downpours that erode the landscape and then sink into the ground. Those downpours would provide groundwater that could be accessed through deep wells and that would perhaps, in depressions in the ground, give rise to the occasional oasis. The Sahara has significant quantities of groundwater. Some of Tatooine's rain and groundwater would then evaporate into the atmosphere, continuing the cycle.

It's very expensive to pump water up from signficiant depths, so perhaps farming the moisture from air would be cheaper and easier. Dr. Condie points out that "In some areas on Earth you have a hard time getting water out of the ground. The water table may be low or nonexistent. Yet in those cases you just go to a well, or pipe in the water from elsewhere." If water is this difficult to pump everywhere on Tatooine, though, and condensation has somehow been made easy, that could be a better solution. If many farmers are doing it, this could lead to a significant depletion of the slight humidity in the air. Luckily, Tatooine seems only sparsely populated.

I said earlier that positing a wetter history for Tatooine was helpful for more than just explaining erosion. If the theory that life on Earth originated near underwater hydrothermal vents is true, then it's hard to imagine how life began on Tatooine. We can imagine primitive underground life developing, as scientists think might exist on Mars, but what about more complex life? Dr. Jakosky says, "If you're going to get big organisms, I think you need standing bodies of water. Microbes can live in the pore space of rocks in the crust. But anything bigger needs room to move." So if Tatooine had more water on its surface at one time, that would make it easier to understand how complex life might have evolved. What kind of life? You'll have to wait for the next chapter to find out.

Twenty years ago, most scientists would have said that planets are probably rare, habitable planets very rare, and habitable planets with all the ingredients for life extremely extremely rare. Now, planets appear to be quite common in the universe, habitable planets a fair fraction of these, and the ingredients necessary for life widespread. Life may be developing, living, and dying in star systems all around us. Although we might need some technological assistance when we visit, with a little help we should be able to survive on a wide variety of worlds. So in this huge population explosion, exactly what sort of neighbors might we find?

2

ALIENS

You'll never find a more wretched hive of scum and villainy.

—Obi-Wan Kenobi, *A New Hope*

A black, triangular-headed alien with glimmering gold eyes pops up in the local cantina. An ancient, eyeless slug lurks within an asteroid. A sinuous snakelike creature with a periscope for a head prowls the Death Star's garbage masher for tasty treats.

One of the most delightful aspects of *Star Wars* is the constant appearance of bizarre forms of alien life. Almost anywhere you go in "a galaxy far, far away," alien life is there. You'll either land in it, step on it, or get eaten by it. The skeleton of a Sandsnake on a Tatooine sand dune creates an echo of the past. A dragon-like creature lurks in an underwater cave on Naboo. Artoo falls into the swamps of Dagobah and is almost instantly eaten, and almost as quickly spit out. Wherever you go, whatever you do, there will always be an alien there to do it with you.

When Luke Skywalker entered that cantina twenty-two years ago, the way we thought of aliens and the way they were treated in science fiction changed forever. No longer were we expected to gawk in fascination at a single alien species; the universe, according to George Lucas, is filled with life. Not only do many planets develop life, but on any one planet, many different species evolve, just as on Earth. Even in environments as inhospitable as Tatooine, Hoth, or an asteroid, life finds a way to survive.

But what sort of life would develop in these various environments? If alien life is indeed plentiful, as recent scientific discoveries lead us to believe, will it look anything like *Star Wars* aliens? Is the universe likely to be home to glowing-eyed Jawas, wriggling Hutts, cuddly Ewoks, hungry Sarlaccs, gooey Mynoks, and goofy Gungans?

HOW ALIEN ARE ALIENS?

On Earth, we're presented with a huge variety of life—organisms with leaves, wings, trunks, claws, flippers, tentacles, hooks, antennae, horns, quills, scales, fur, fangs, shells, slime. Organisms that reproduce through spores, seeds, division, sex with another, sex with themselves. Organisms that live inside other organisms; organisms that live attached to the outside of other organisms; organisms that live in water, air, rock, dirt, blood, ooze. Over the history of our planet, tens of billions of different species have existed. Dr. Pickover says, "When I gaze upon crazy-looking crustaceans; squishy-tentacled jellyfish; grotesque, hermaphroditic worms; and slime molds more alien than the wildest dreams of science fiction writers, I know that God has a sense of humor, and we will see this reflected in other forms in the universe."

Considering that such a variety of creatures developed on our single planet, it seems unlikely that aliens will happen to have human characteristics or form. With two arms used to perform tasks, two legs used to walk, and a head with sensory apparatus, many *Star Wars* aliens have the general form of a human: Yoda, the Jawas, the Sand People, Jar Jar Binks, the Wampa ice creature, Greedo, Admiral Ackbar, the Ugnaughts of Cloud City, the lizard bounty hunter Bossk, the tentacle-headed Bib Fortuna, singer Sy Snootles, Lando's copilot Nein Numb, Chewbacca, the Ewoks, the cantina band, and many other miscellaneous aliens in the movies.

Yet *Star Wars* provides us with many non-humanoid aliens as well, offering a wider variety of aliens than any other science fiction movies, with Banthas, Tauntauns, Sandsnakes, Sarlaccs,

Hutts, Dewbacks, Mynoks, space slugs, and many more. These make the *Star Wars* universe feel real, vibrant, and unique. Dr. Michio Kaku, Henry Semat Professor of Theoretical Physics at the City University of New York and author of *Hyperspace* and *Visions*, agrees. "In *Star Wars*, the aliens don't look like us anymore. They tried to have aliens with different architectures. In that sense, *Star Wars* is more realistic than some of the stuff I've seen." But is this what alien life would really be like?

The most likely alien life we'll encounter will resemble terrestrial bacteria. Bacteria developed first on Earth about 3.85 billion years ago, and remained the sole type of life for billions of years. Multicellular organisms didn't appear until about one billion years ago, and animals evolved only six hundred million years ago, quite recently in Earth's life span. Even though complex life-forms are now common on Earth, bacteria still make up the majority of life on our planet. Dr. Stephens believes we're going to find the same ratio of primitive to more sophisticated life everywhere in the universe. "I think what we're going to find is enormous numbers of planets and moons where the first steps of life have occurred, even up to the formation of things that resemble bacteria." So we'd likely encounter many planets with primitive life-forms, and a much smaller number with more complex life. Dr. Jakosky comes to the same conclusion. "It took three billion years on Earth to go from single-celled organisms to multicellular complex organisms. That means that's not a likely event. Once we got complex organisms one billion years ago, though, Earth experienced a rapid explosion of diversity of life. Once complex life develops, tremendous diversity is likely to arise." What would this complex life be like?

To consider what alien life might look like, it's helpful to think about how terrestrial life came to look like it does. Species are created through evolution. Random genetic mutations occur as organisms live and reproduce. If a mutation happens to be favorable, helping the organism to survive in its particular surroundings, it's more likely to reproduce and pass that mutation on to offspring. If the mutation happens to be unfavorable, the organism may die before it has a chance to reproduce, and so the mutation will disappear from the gene pool.

Evolution involves a lot of chance circumstances, both genetically and ecologically. Dr. Jakosky calls these chance circumstances "accidents of history." A mutation could occur that makes a fish much more fit for life in the tiny pond where it lives. Yet it happens to be born in a dry year, and before it has a chance to reproduce, the pond dries up and the fish dies, so the mutation is never passed on. At the same time, another fish is born that has a mutation that makes it much less fit for life in the tiny pond. Normally, the first fish would eat this one for breakfast. Yet as the pond dries up, the second fish finds it can survive for brief periods outside water, long enough, in fact, to flop over to a larger lake a few feet away that doesn't dry up. This odd fish, which can move between land and water, survives and reproduces, and perhaps becomes the ancestor of all vertebrates, including man. If instead of a dry year it had been a year of heavy rains, that fish would never have survived, and quite different creatures might have evolved.

Dr. Jack Cohen, reproductive biologist and consultant for the Mathematics and Ecosystems Departments at the University of Warwick, points out some interesting characteristics of the fish that came out of the water three hundred million years ago to become the ancestor of all vertebrates. "It had its airway crossing its foodway, and it had a reproductory system mixed with its digestive system. There were many fish that didn't have those mistakes, and one of those could have crawled out of the water." Chance, then, has played a large role in our development, making us creatures that breathe and eat through the same opening, our mouth, and reproduce through organs intertwined with our excretory system. Many *Star Wars* aliens seem to share these mistakes—at least they appear to use the same opening to breathe and eat, which seems quite odd. We can only speculate on their reproductory systems.

How these "accidents of history" might combine on another planet is very difficult to predict. Dr. Tim White, professor of integrative biology at the University of California at Berkeley, believes that "it's hard for us to imagine other kinds of life. Animals adapt, through evolution, to their environment, and we don't know about extraterrestrial environments outside our solar system. In addition, you have chance events that end up knocking things off balance every once in a while, and thereby structuring the fabric of life and making its evolution unpredictable."

Even a planet similar to Earth would most likely give rise to significantly different organisms. In fact, scientists believe humans wouldn't even have evolved here unless the dinosaurs were killed off by a huge meteoroid sixty-five million years ago. Dr. White says, "You take away this event or that event, and we're not having this conversation."

Most likely, aliens are going to be stranger than we can imagine. Dr. Pickover agrees. "Considering that octopi, sea cucumbers, tube worms, and pine trees are all very closely related to us, an alien would look less like us than does a squid."

So is there no way to tell what aliens might look like? Well, aliens are life-forms, like us, and face some of the same problems we face—problems of movement, sensing and manipulating their environment, nourishing themselves, and reproducing. Solutions to these problems that are valid on Earth ought to be equally valid in "a galaxy far, far away." Thus, even though alien life would have taken a very different evolutionary course and would have DNA quite different than ours—if it has DNA at all—it might end up with some similar characteristics.

Life on planets has to deal with gravity, which will probably lead to organisms with a specific top and bottom. Sophisticated life that moves will probably have a front and a back. To sense their environment, organisms need one or more of the following abilities: to detect a useful spectrum of electromagnetic radiation (to see); to detect changes in the surrounding atmosphere (to hear and smell); to detect heat and evaluate surfaces (to touch); and to evaluate food (to taste). If an organism must seek out food, then it would make sense to concentrate these sensory organs at the front end of the organism. To manipulate the environment, sophisticated organisms require appendages. To move through that environment, they need some method of locomotion. Some sort of symmetry will make mobility easier. But how likely is it that a creature might move using two limbs like a man, four limbs like a giraffe, a muscular trunk like a snake or a fish, wings like a bird or an insect, a tentacle like an octopus, or some other method?

We can get some idea of how common or likely a certain solution might be by examining how many times that trait independently developed on Earth. Solutions that developed in different times and places must be particularly useful or efficient. For example, flight developed three separate times on Earth: once with birds, once with insects, and again with bats. Eyes developed four separate times. So these solutions may be a bit more likely to develop on another planet. Dr. Pickover points to three quite unrelated animals: a dolphin, which is a mammal; a salmon, which is a fish; and an ichthyosaur, which is an extinct reptile. "They all swam in coastal waters darting about in search of small fish to eat. These creatures have very little to do with one another biochemically, genetically, or evolutionarily, yet they all have a similar look. They are nothing more than living, breathing torpedoes. They have evolved streamlined bodies to help them quickly travel through the water. We might expect aquatic aliens that feed on smaller, quick-moving prey to also have streamlined bodies." Other solutions have occurred only once on Earth, making them seem less likely to occur on another planet. For example, while all land animals have developed methods of acquiring water, only one, the elephant, uses a long trunk.

Dr. Stephens believes that the form these solutions take is largely determined by physical constraints, which in turn are consequences of the laws of physics, chemistry, and biology. For example, let's consider fingers or, more generally, digits. If aliens arrive here in spaceships, how many digits are they likely to have? An intelligent, space-faring race needs to be able to manipulate its surroundings with limbs, and digits at the end of those limbs. Dolphins may be intelligent, but they'll never light a fire and never build a spaceship. The realities of the physical world make certain numbers of digits more functional than others. "Is there an ideal number?" Dr. Stephens asks. "Clearly the answer is yes."

One digit is obviously not going to be terribly useful at performing complex tasks. Similarly, two digits aren't very good, which you've discovered if you've ever tried to do anything delicate wearing mittens. Dr. Stephens says, "Three is some kind of minimum threshold if you're going to do serious manipulations." If three is the minimum, what is the maximum? "When you get up to seven or eight, I think you have a difficult time accommodating that many digits on the end of an extremity and things would start becoming awkward." While we haven't done experiments that might prove five is the ideal number, Dr. Stephens believes five likely is the ideal. "What we've learned recently is that our amphibian ancestors did not have five digits but had six or seven digits. Evolution and selection chose five for us, suggesting five is some sort of ideal." Dr. Stephens then believes that intelligent aliens would likely have five digits, as we do.

Since certain physical constraints are valid throughout the universe, Dr. Stephens concludes that even though intelligent aliens might have evolved from vastly different organisms than we did, they may very well be humanoid. "The probability of finding an alien that looks like us is perhaps as high as 80 percent."

Yet he seems to be in the minority. Dr. Cohen suggests the opposite is true. "Finding another planet with our kind of dinosaurs or people is more unlikely than finding a remote Pacific island on which the natives speak perfect German." Aliens would have gone through an entirely different evolutionary procedure, suffering different "accidents of history," which would have led to different adaptations. Dr. Pickover considers humanoid aliens "far-fetched. Some of the *Star Wars* creatures seem a little too human looking considering the quite different evolutionary pathways we'd expect." For Dr. White, the most troubling characteristic of intelligent aliens is that most of them are bipedal. "There is no necessary correlation between bipedality and high intelligence." Dr. Kaku believes there are only three key elements an intelligent alien must have. "An opposable thumb or tentacle of some sort, language to communicate, and stereo eyes to hunt and strategize. Other than that, all bets are off." Dr. Jakosky agrees. "There's no intrinsic evolutionary drive toward a human shape, even though we can make all these arguments about how wonderful we are."

While the number of aliens with human characteristics may be high, *Star Wars* aliens embody many nonhuman characteristics as well. Do the aliens make use of more "common" solutions, which would make them more likely aliens; do they use solutions that occurred only once on Earth, making them seem unlikely to occur ever again; or do they use uniquely alien solutions?

FUZZBALL OR GENIUS?

If we do find alien life, will it be intelligent? *Star Wars* abounds with intelligent aliens, from Chewbacca to Jabba to Greedo to Jar Jar to Yoda. When I speak of intelligence here, I mean intelligence comparable to the sophisticated self-conscious intelligence humans have. My iguana, Igmoe, is extremely intelligent, yet even I won't claim that he has the powers of thought, reasoning, and understanding that we have—at least, he fails to appreciate *Star Wars*, which is a failing of intelligence in my opinion.

Some scientists believe that while alien life may be plentiful, intelligent alien life is much less probable. They argue that while evolution works through "survival of the fittest," being smarter does not always make you more fit. Organisms can be very successful in surviving and reproducing without high intelligence. Cockroaches, for example, are an extremely hardy species and may well outlive man. If additional intelligence was an advantage for them, so the argument goes, then we should have seen them growing progressively more intelligent over the generations, and we haven't. Of course, I don't know anyone who's administered an intelligence test to a cockroach. Dr. Jakosky agrees that evolution doesn't necessarily encourage intelligence. "Organisms tend to have a brain just big enough to handle their body and not a lot extra. We're the exception." He points out that in four billion years and tens of billions of species, only man has developed self-conscious intelligence, "so that says it's an uncommon event." Or in other words, it's an uncommon solution to the problem of survival. Dr. Pickover sees our intelligence as an evolutionary accident. "I believe that alien life will be unintelligent and unable to build crafts to leave their world. This is the prime reason why the universe of *Star Wars*, which thrives with intelligent life, is unrealistic."

Yet other scientists believe intelligent life may be more common. While intelligence within a particular species may not increase, scientists point out that as life has evolved on Earth into more complex, advanced forms, the size of the brain relative to the body has increased, and intelligence has increased. And while human-level intelligence may have only developed once on this planet, lower levels of intelligence developed independently in all the different classes of animals. Astronomer Carl Sagan argued that more intelligence is beneficial to any organism, helping it find food and cope with changes in its environment. In that case, intelligent organisms would have a better chance of survival. Thus intelligence would be a natural consequence of life and evolution. Dr. Frank Drake, head of the SETI Institute, estimates ten thousand to a hundred thousand intelligent civilizations exist in our galaxy.

But if intelligent alien life is common, then why haven't any aliens dropped in on us? This question is known as Fermi's paradox, and scientists have struggled for an explanation for more than two hundred years. Some of the explanations they've come up with?

- Alien civilizations self-destruct before they develop interstellar travel.
- Aliens do visit us, but they like to keep their visits secret. Word could get back to a pesky bounty hunter.
- The stars are so far from each other, visits are just too difficult and expensive. Maybe if we kicked in some money for gas. . . .
- Earth is just too far from the bright center of the universe to draw alien tourists.
- Aliens prefer to stay home, put on a pot of stew, and observe us through the Force.

DO YOUR EARS HANG LOW?

Of all the alien species we meet in *Star Wars*, we know the native planets of only a few. Without this information it's difficult to say whether their traits are likely or not. Even with this information, it's very difficult to draw any firm conclusions. For example, say giraffes are not a species on Earth but an alien species in the next *Star Wars* movie. We could speculate endlessly about the viability of such an alien and the bizarre environment that might have spawned such a creature. That long, heavy neck? How unlikely! It would fall on its face. It must come from a planet of light gravity that allows it to survive with such skinny stick legs. The planet must be covered with trees that hold their leaves an enticing fifteen or twenty feet off the ground. These conclusions seem reasonable, yet every one is wrong. Thus, any speculation about specific aliens can be little more than educated guesses at this point.

One factor may help a bit. Since the aliens we tend to see in the movies are those that can survive in the same environments as humans, we can assume that they probably developed in environments not radically different from Earth. Now let's examine some of our favorite aliens.

The Phantom Menace introduces the lovable goofball Jar Jar Binks, a native of Naboo. Jar Jar has the basic exterior characteristics discussed earlier for complex, intelligent life: a top and bottom, a front and back, bilateral symmetry, sensory organs, a method of locomotion, and appendages to manipulate the environment. Generally humanoid in shape, Jar Jar has a number of distinctive characteristics that actually make him a fascinating mystery. Clearly the "accidents of history" that guided the evolution of Jar Jar's species were much different than ours, but led to a body form that generally has much in common with a human's. His long mũzzle gives him the look of a camel or horse, his long ears are a bit rabbitlike, and the patterning of pigment on his arms is reminiscent of a lizard.

Jar Jar's most prominent trait—literally—is his eyes. Popping out from the top of his head like two ears with an identity crisis, they are unlike the eyes of any Earthly creature. The unusual location of the eyes and their structure suggest that their position may be critical to the survival of Jar Jar's species, the Gungans. While we don't know exactly where on Naboo the Gungans evolved or how, we do have terrestrial animals with some similarities that may illuminate Jar Jar's situation.

In the crocodile, the nostril openings and the eyes are the highest parts of the head. This allows the crocodile to float on the surface of the water, with most of his body and head submerged, and yet see and smell his surroundings. Thus the crocodile can wait, looking like a floating log, until a potential meal comes to the edge of the water for its last drink. Then with a quick snap, it's mealtime. Perhaps the Gungans, or their ancestors, obtained their food by some similar method. Their nostrils are positioned high on their heads, though not as high as the eyes.

Another terrestrial model offers a different possibility. Many crustaceans, as well as some insects and fish, have eyes on stalks. Eyestalks allow the eyes a great freedom of movement. While Jar Jar's eyes are not on stalks and so can't have the level of mobility of stalk eyes, his eyes still probably have some mobility, with muscles at the base of the eyes allowing them to tilt forward or back or twist to one side or the other. In some creatures with stalk eyes, each eye is at the end of a rod of cartilage. At the base of the rod is a muscle that controls the movement of the rod. Thus the rod can be moved around, aiming the eye in virtually any direction. Since crustaceans tend to be slow moving and are slowed further by the viscous water, they can't dart their heads about quickly like birds to monitor their surroundings. Jar Jar can move quickly, though. So it's unclear what advantage moving eyes would have when he could move his head just as fast.

Some mobility in his eyes, though, can potentially provide a wide field of view. If a creature is an herbivore, a wide field of vision is helpful to keep watch for predators sneaking up on it. In terrestrial creatures, most herbivores get a nearly 360-degree view by having eyes that face to the side. If a creature is a carnivore, stereoscopic depth perception is helpful to precisely target and seize its prey. Carnivores then tend to have both eyes facing ahead. Jar Jar's eyes do seem to face ahead when at rest, so he may have the best of both worlds: stereoscopic depth perception and a wide field of vision when needed.

Another advantage of this construction is with the eyes removed from the skull, there is more room for the brain. In many terrestrial species, brains and eyes must compete for skull space. The larger the eye, the greater the visual resolution. So larger eyes with sharper vision would be helpful to all creatures to survive. Yet the bigger the eyes, the less room is left for the brain. Birds, whose high-speed flying requires sharp vision, give up more skull space to their eyes than their brain. Removing the eyes from the skull would allow Jar Jar to have a bigger brain than a camel or horse.

This solution to the skull-space problem, though, makes the eyes vulnerable to damage, rather than protected within bony sockets. Dr. Stephens points out, "If you look at any vertebrate organism, the senses are quite well protected by bone. Even the crocodile has a ridge of bone on top of its eyes. No vertebrate organism I can think of has any sensory organ hanging out there in the breeze without protection." A fall on the head could get squishy, and a predator could tear Jar Jar's tasty nuggets off in one bite. It seems the eyes must offer a significant advantage to make up for this danger.

His eyes yield yet further information. Animals that live in nocturnal environments must develop very sensitive eyes, eyes that have both very large pupils to allow in as much light as possible and specially designed retinas to detect the dimmest light. The light-sensitive retina at the back of the eye is normally made up of two kinds of cells, rods and cones. The rods are sensitive to dim light, while the cones handle bright light and provide color vision. Nocturnal animals have more rods than cones, which gives them better night vision. But when these animals also go out into bright light, they have trouble coping.

Their pupils can contract to shut out most of the blinding light, muscles in the eye tightening to make that opening smaller. But the specific arrangement of the muscles sets a limit on how small a round pupil can become. Even at its smallest, it may let in more light than a nocturnal creature can handle. A slit pupil, on the other hand, has its muscles arranged differently. When the pupil is dilated, it appears round, yet when it contracts, the muscles can close it down to a narrow slit, allowing only a tiny fraction of light to enter. The slit pupil is much better for nocturnal animals that also venture out in daylight. The crocodile, which is mainly nocturnal but also enjoys basking in sunlight, has a slit pupil. We can theorize that Jar Jar is mainly nocturnal as well. That must make it hard for him to adopt the diurnal schedule of his human companions. Qui-Gon may find him prowling around at night, like you'd find your pet cat.

Another clue to his lifestyle may be in his rather long, flexible neck. Many terrestrial animals have developed long necks, so it seems a fairly reasonable trait to find in an alien. Terrestrial animals use long necks for a variety of reasons. Giraffes use them to reach leaves high off the ground; storks use them to catch fish. We have a problem with any theory that has him using his neck to get food, though. A biped with two legs and two arms, like Jar Jar, would be much more likely to catch a fish with his hands, or to reach up with a hand and pull a branch with some tasty leaves down to mouth level. Dr. Stephens asks, "Why would you develop a long neck if your grasp is twice as high as your head?"

Perhaps, if he is operating like a crocodile, he waits until prey come close to his mouth and snaps them in, the long neck giving him additional mobility. Or perhaps the advantage of the neck is not to reach food, but to elevate the eyes even further. Camels have both long legs and long necks, which together serve to elevate their heads above the blinding, choking particles of desert sandstorms. A similar environmental condition could explain why Jar Jar's body works so strongly to elevate his eyes. Perhaps a layer of ground fog tends to form during the night in the region where the Gungans evolved. This occurs in different areas on Earth, such as river valleys, wetlands, or in coastal regions. The fog may tend to dissipate at a certain height, offering an advantage to those whose eyes are above the obscuring fog.

Another interesting trait is Jar Jar's long muzzle, which on Earth is common to quadrupeds. Quadrupeds tend to have long muzzles, since they need to use their mouths to hold and manipulate food as they eat it. Bipeds with arms, like humans, can use their hands to hold and manipulate food, so they don't need a long muzzle. Jar Jar's muzzle leads Dr. Stephens to "expect him to be very clumsy with his hands." Jar Jar definitely has some coordination problems.

One final clue to Jar Jar's nature is his huge, floppy ears. Large ears in terrestrial animals help animals radiate excess heat into the surrounding air. Desert-dwelling rabbits and other desert animals often have extra-large ears to help cool their blood. The large ears of African elephants serve the same purpose. Prominent veins in the ears help bring the blood close to the skin's surface for cooling. In addition, the elephant can flap his ears to speed cooling. Jar Jar's ears may work in a similar way. They may also serve as communication devices. Elephants move their ears to assert their dominance over other males and to defend their territory. Perhaps the movies will give us further clues as to whether Jar Jar's ears function in this way as well.

We can thus theorize that the Gungans may have evolved in a hot climate where threatening predators lived. Hunting mainly at night offered one method for avoiding the heat, while their huge ears offered a method for dealing with heat when they come out during the day. Environmental conditions may have existed that made it difficult to see near ground level, so that elevating their eyes offered great advantages. Or the placement of their eyes may have helped them surprise prey. Yet the eyes are Jar Jar's Achilles' heel, and he must guard them at all times. One good tumble, and it's lights out.

THE DAWN OF WOOKIEE

Of all *Star Wars* aliens, the one I feel closest to is Chewbacca. Over seven feet tall and more than two hundred years old, Chewie is a hulking walking carpet, loyal, courageous, at times fierce, at times—when his nose is telling him things no one else seems to notice—a bit skittish. According to the *Star Wars Encyclopedia*, Wookiees come from the forest planet of Kashyyyk, where they build cities a mile off the ground in the tops of massive trees, rather like Ewoks. Their long limbs would help them climb trees and swing from branch to branch. Claws on their feet, which are narrow and sharp, seem potentially helpful for climbing, though Chewie's are clipped short—or perhaps they're just retracted, like a cat's. We can't see if Chewie has claws on his hands, since, like most of his body, they're shrouded in fur. Wookiees are very strong and can rip people's arms out of their sockets, as Threepio well knows. They communicate through a language of cries, howls, and grunts. While they can understand English, they are unable to speak it.

Since from outward appearances Wookiees are basically humans covered in fur, they seem like viable organisms, in which all the pieces could fit and work together. Since body coverings like fur, hair, and feathers have developed independently several times on Earth, fur on an alien seems reasonable. Such body coverings help insulate a creature from external temperature changes, so Wookiees most likely face temperature fluctuations similar to those on Earth and use a similar solution to cope with them. Chewie's sharp teeth, front-facing eyes, and claws all point toward a carnivore. Chewie's sensitive sense of smell would aid in tracking and locating prey.

Humans hunt prey as well, though—at least our ancestors did—so we might question why Wookiees have a better sense of smell than we do. Scientists have recently found that 72 percent of the genes that govern the formation of our olfactory receptors, the proteins that enable us to smell, are mutated so badly that the receptors don't work. Our sense of smell has deteriorated greatly. Exactly when these mutations occurred is not yet known; they may have occurred during human evolution, or even earlier, before man developed. Apparently a high-powered sense of smell is not critical to man's survival, and so those with less effective noses still survive.

In Wookiees, smell must play a more critical role. They may need—or have needed—smell to aid in the tracking of prey, which could hide in the dense forest. Or they may use their sense of smell to detect the emotions of other creatures. An angry snake, for example, has the distinctive smell of a wet dog, and knowing when a snake is angry could come in very handy. Wookiees may also use smell to send signals to each other, marking territory, indicating a desire to mate, or warning of danger. Since the vocabulary of Wookiees may be limited, scents could offer another method of communication. A dog's sense of smell is one million times more sensitive than ours. If Chewie's is equal to or even greater than this, he may be getting a wealth of information through his nose. He could tell whether Jabba is hiding bounty hunters with blasters in the next room, a caseload of spices, or a frozen Han Solo.

A strong sense of smell could even explain why Wookiees seem rather emotional. Scientists believe smell, more than any other sense, evokes strong feelings. The smell of the sweater of a loved one calls up a vivid sense of that person, triggering strong memories and feelings; the smell of fresh-baked chocolate-chip cookies triggers pleasure and excitement (in me, anyway). Odors can be informative, disgusting, delightful, frightening, and evocative.

Scientists believe that smell was one of the first senses to develop in terrestrial life. While information from our other sensory organs is relayed to the neocortex, the part of the brain involved in higher thought, scent information is passed directly to the most primitive part of the brain, the limbic system, and particularly an ancient structure called the amygdala. The amygdala is involved in storing emotional memories and can imbue events with intense emotions.

In our primitive ancestors, a smell triggered a strong emotion that in turn triggered a behavior: good food—eat! Today, while reason may keep us from acting on our desires—keep your hands out of the cookie jar—scents retain their direct line to our emotions. These brain connections are largely the result of evolutionary "accidents of history," yet perhaps Chewie's senses are wired in a similar way. With his more refined and acute sense of smell, scents may call up powerful emotions in him.

While the Wookiees' ability to smell is much greater than ours, their ability to speak seems much more limited. In this way, Wookiees are similar to chimpanzees. Chimpanzees communicate by expressions, gestures, and many distinct vocalizations, including screams, roars, hoots, and grunts. They lack the ability to articulate the variety of sounds man can.

Many elements contribute to man's ability, including the lips, tongue, teeth, and the hard and soft palates on the roof of the mouth. The most critical element is the larynx, where the vocal cords are located. In humans, the larynx is lower in the throat than in apes. This change occurred as man began to walk erect, his brain size increased, and the location of the skull's fastening to the spinal cord shifted to better balance the head. The lower position of the larynx creates a tubular cavity in which sound can resonate. Our relatively low-pitched speech arises from this cavity. The structure of the chimpanzee's larynx doesn't allow it to make many of the sounds needed for human speech. But they have been taught to say a few limited words, such as "mama"—just like in *Planet of the Apes*. Wookiees, then, may not have their larynxes positioned in a way that allows them to reproduce human speech. And why should they? It seems much more believable that aliens wouldn't be set up with the exact structure needed to reproduce human sounds.

With his great number of similarities to us, Chewie doesn't pose as difficult a puzzle as Jar Jar. Yet he does raise one compelling question: Why are Wookiees bipeds? Bipedality has evolved a number of different times on Earth, in dinosaurs, kangaroos, birds, and hominids. So it seems a trait we might possibly find in aliens. Yet is it a trait we would likely find in Wookiees?

We don't know if the Wookiees always lived in the trees, or if they only moved there recently, once their level of technology allowed them to construct cities in the treetops. Yet it seems odd that a species that evolved on the ground would suddenly decide to move up into the trees. They would feel more at home on solid ground, the animals that served as food would be on the ground, and the conditions most favorable to their body structure and life-style would be on the ground. Climbing a mile down to the ground to hunt for dinner and a mile back up would make hunting a diffi-cult proposition.

It seems more likely, then, that Wookiees always lived in the trees, and that they're able to satisfy all their needs without climb-ing down to the surface. If the Wookiees did evolve in the trees, then, as seems more likely, we might wonder why they'd develop a bipedal gait. Walking upright doesn't seem the easiest way to get around on a tree. Even if you could balance upright on a gigantic tree branch, why risk it? Quadruped tree dwellers can move with great agility and speed.

Dr. White would prefer to have more information before specu-lating about the evolution of the Wookiee. "In the unlikely event that I was presented with a real, rather than imaginative, Wookiee, I'd like to check out its living relatives and its ancestors, as much as we could figure out from the Wookiee fossil record, in order to explain where that creature came from." In the absence of that information, though, can we learn anything from comparing Woo-kiees with Earth's tree dwellers?

A four-limbed creature needs many specific characteristics to walk on two limbs. The bones of the vertebral column, pelvis, leg, and foot need to have the proportions and shapes to withstand the stresses placed on them and allow easy movement. The muscles of the trunk and thighs need to develop the ability to balance and support the body's weight and to propel it forward.

Scientists believe that a tree-climbing life helped prepare man's ancestors to walk on two legs. Tree climbers tend to develop front legs somewhat different than their back legs, the front limbs reaching for food or branches, and the back legs supporting the weight of the body. Anthropologist Michael Seaman at Yale Uni-versity points out that "Since they're climbing up and down all the time, they hold their torsos erect." These adaptations actually serve as the preliminary changes necessary to prepare for a biped gait. So climbing creatures that evolve into bipeds seem reason-able.

Yet the accepted theory for many years has been that a key element in the development of the biped gait in man's ancestors, the hominids, was a change in climate. About four million years ago, the climate on Earth began to get drier. East Africa, which had been a moist woodland, began to change into an open grass- and scrub-covered savanna, the trees dying off. Man's ancestors then had to leave the trees and adapt to life on the ground. In fact, footprints preserved in volcanic ash reveal bipedal locomotion had developed in this region 3.6 million years ago.

If the savanna theory is true, and bipeds did develop because a climactic change led hominids to move from the trees to the ground, one might then wonder how Wookiees—and Ewoks—ever became bipeds. Since they both are said to live in trees on heavily forested planets, and we've theorized that they evolved in these trees, the conditions seem incompatible with the savanna theory.

In the last five years, though, the savanna theory has been challenged by new fossil discoveries and revised estimates of the climate during the development of hominids. It's now believed that Africa's climate did not become particularly arid until 2.8 million years ago, while we have recently discovered signs of bipe- dal locomotion as early as 4.2 million years ago. The recent dis- coveries of two previously unknown hominid species older than any before found are adding to the uncertainty. There are even indications that bipedalism may have developed independently more than once in different species of hominids, which would make it a somewhat more likely trait for aliens. While the savanna theory has been thrown into serious doubt, the problem is that scientists have no clear theory to replace it.

The leading contender appears to be one proposed by Dr. Owen Lovejoy of Kent State University. In his theory, the stimulus for bipedalism was the pairing off of male and female hominids into monogamous couples. The male began to provide food to the female and babies. When hominids developed this new strategy, it allowed females to give birth more frequently and males to gain exclusive sexual access to a female, which allowed successful pairs to pass their genes on to a large number of offspring. What does sex have to do with how you walk? Well, with this new lifestyle, the male needed to carry large amounts of food back to the female and children. To do this, he needed to walk upright and free his arms for carrying. Dr. White believes "the Lovejoy model is the best available model."

If Dr. Lovejoy's theory is true, it may reveal how Wookiees developed bipedal strides. While we don't see much of the Wookiees' lifestyle in the movies, key information is provided in the 1978 *Star Wars Holiday Special*, a two-hour TV show that is so awful it's fascinating. Yet the show provides exactly the information we need. In this show, Chewbacca visits Kashyyyk, returning to the home where his wife Malla and son Lumpy live. And so we see the monogamous lifestyle that may offer a possible explanation of the Wookiees' biped gait. My guess is that plenty of males would be willing to try walking upright on a tree branch if it meant they could have sex on a regular basis.

JUST BECAUSE IT GOES "HO HO HO" DOESN'T MEAN IT'S SANTA

The award for most disgusting alien would have to go to Jabba the Hutt. Over sixteen feet long, with a bloated sluglike body, stubby arms, and a head like a giant pimple, Jabba slithers his way through the galactic underworld, with his home base a great palace on Tatooine. According to the *Star Wars Encyclopedia*, Jabba secretes mucus and sweat through his skin, making him slimy and slippery to catch by enemies. "His high exaltedness" enjoys snacking on marine life, which he keeps in a small aquarium beside the dais in his palace's throne room. This suggests Jabba's native habitat may be near the water.

Although terrestrial slugs are only a few inches long at most, they offer the closest comparison we have to Jabba. Slugs have soft, slimy bodies and tend to be nocturnal. While common slugs eat fungi and decaying leaves, some slugs are carnivorous, like Jabba, eating snails and earthworms. Slugs are hermaphrodites, containing both male and female sexual organs. The *Encyclopedia* tells us Jabba is the same. We should, in fact, call Jabba *it*, but that seems strange, so let's go along with the *Encyclopedia* and call Jabba *him*. A slug can reproduce by taking on the male role and injecting sperm into another, or by simply combining its own sperm and eggs, in essence having sex with itself. Jabba may have both of these options as well, which makes one wonder why he would be attracted to the scantily clad Leia. Dr. Pickover points out, "The chance of Jabba finding a human female alluring is about as great as you and I finding a female squid alluring." I prefer a male iguana myself.

The belly of a slug is actually a single, tapered foot. The slug moves by generating waves of muscular contraction that ripple down the foot from the tail end to the front end. Also helping the slug along are tiny cilia on its foot, and a mucous sheet secreted by the front of the foot that provides a slimy carpet to help the slug glide ahead.

This secretion of mucous causes the slug to lose a great amount of water, so all slugs need a moist environment to thrive. Slugs and snails are both in the class Gastropoda, but snails developed in areas with irregular moisture, their shells a safe place to withdraw and contain their moisture in dry times. They can literally seal themselves in and wait years for rain. Slugs, without a protective shell, developed in areas that are moist year-round. To make sure they don't run out of water, they conserve what they have, reabsorbing the water in their urine. But if slugs require moisture, how can Jabba survive on the desert planet of Tatooine?

In neither the special edition of A New Hope nor The Return of the Jedi do we see Jabba outside. Even as he is ordering Han Solo to his death in the Sarlacc, he remains below decks in his barge, without a clear view of the spectacle. I can't believe Jabba would willingly miss seeing Han's death. Jabba must need to remain in darkness most of the time, sheltered from the sun, the heat, and the dry air. Jabba's preference for the dark goes along with a slug's nocturnal lifestyle. Jabba's slit-shaped pupils support the theory that Hutts are naturally nocturnal. Both his barge and his palace are kept dark, and he may have humidifiers of some kind keeping the air humid. Of course such a method would be extremely costly on a desert planet where water is scarce. But he is Jabba the Hutt. And Tatooine, isolated and unwatched, may offer a very comfortable home for a crime lord.

Jabba does seem able to get around, at least in a minimal way. In A New Hope, we see him in the Millennium Falcon's docking bay, which certainly is not equipped with any special humidifiers. Here, Jabba's size may give him an advantage over terrestrial slugs. A large slug will dry out more slowly than a small slug. So Jabba could survive short periods of low humidity without any problems.

The comparison to a slug does suggest several ways in which you might cope with a Hutt, if you find yourself face to face with one and don't have a chain to strangle him with. First of all, don't try to grab him, because the slime will help him slip away. And if you do manage to get hold of him, the Hutt probably has another trick up his sleeve. Several kinds of land slugs have the ability to break off the back portion of their foot. This portion twitches violently, distracting the enemy, while the rest of the slug slides away. In *A New Hope*, Jabba seems shocked when Han jumps onto his back. I wonder if he was nearly startled into breaking his body apart. To kill a Hutt, you could always leave him in the desert to dry out. If you want to speed up the process, you can irritate his skin. Just as getting an irritating speck of dust in your eye makes your eye water, getting some irritating material on the skin of a slug makes it secrete huge amounts of slime. Spreading ashes or salt over the ground where a slug must travel works very well. The overproduction of slime dehydrates, exhausts, and finally kills the slug. This could have saved Han, Leia, and Luke lots of trouble.

Is it likely that an alien would look like a slug? The fossil record reveals that from more primitive forms, land-dwelling, carnivorous slugs developed independently several times on Earth. This suggests that these characteristics are fairly useful and efficient, and might possibly arise again on another planet. So when you head out on that space vacation, you may want to take a bucket of ash with you. And keep an eye on the right side of the Hutt's head. The slug keeps its male reproductory organ there.

SLUGFEST

One of the most surprising aliens, and one of the most difficult to understand, is the space slug that lives inside one of the Hoth asteroids. The slug is hidden inside a cave or tunnel in the rock, and Han unknowingly lands the *Millennium Falcon* inside it. The slug seems to use a strategy similar to the Sarlacc on Tatooine, or the terrestrial ant lion or stargazer fish. The slug stays put with its mouth open, waiting for prey to fall in or come close enough to be grabbed. The stargazer fish buries itself tail first in the sand, so only its eyes and mouth are visible—and only visible if you know what to look for. When prey swim by, it gulps them down. Many terrestrial organisms have such lifestyles, which seems to make the space slug fairly believable.

And since we've discussed organisms that live deep underground on Earth, surviving on only rock and water, this suggests a narrow possibility that an organism could find sufficient nutrition on an asteroid. Some asteroids are believed to have a layer of permafrost, and some reveal evidence that in the past they were heated enough so that their interiors melted. While it's extremely unlikely that any heat would have lasted long enough to develop life, we could perhaps believe that under some unusual circumstances, the largest, hottest asteroid could have had some tiny stirring of life within.

Yet the slug is clearly not feeding off of water and rock. It is a predator, hiding in wait with sharp teeth to grab and tear prey apart. This means there must be an entire ecosystem in the asteroids. There must be creatures on whom the slug normally feeds— and I think it's fair to assume they aren't spaceships, since a pilot would "have to be crazy" to fly into an asteroid field. To support something as big as the space slug, these food creatures must be fairly large or fairly plentiful. Even if the slug lies dormant for long periods, it has to feed sometime. Yet we see no sign of organisms. And if there are microscopic organisms living deep within the rock, the slug cannot be feeding on them, since its mouth is pointed toward space. We might posit some other microscopic organisms somehow floating through space, the giant space slug feeding off them like a whale feeds off tiny plankton. But then why would the slug have such fierce teeth, or the ability to lurch out of its burrow and catch prey? The *Millennium Falcon* seems too small for it to even notice. Its prey must be large and active.

Let's put aside for a moment the problem of what it eats. Say there is a large, constantly replenished supply of insane Corellian space jockeys that fly into the asteroid field, and have been flying into it for the last billion years. Could the space slug live in such an unfriendly environment? The slug would face a host of problems, including cold, meteoroids, and high-energy particles. But let's focus on just one: pressure. Asteroids are too small to hold an atmosphere, so the slug is not sheltered in any way from the vacuum of space.

Many people believe that the lack of pressure in space would cause living creatures to explode. The reasoning goes like this. Our bodies are in a state of hydrostatic equilibrium with Earth's atmosphere. At sea level, the atmosphere pushes in on us with a pressure of fifteen pounds on every square inch of our bodies. The fluids and gases inside our bodies exert an equal pressure outward. Any change in the pressure causes difficulties. If you are subjected to pressures less than 14.7 pounds per square inch, the body swells and bubbles of gas form in the blood. Divers go through some of these stresses when they move too quickly from high-pressure ocean depths to the lower-pressure surface, suffering the "bends." If you quickly bring a deep-sea fish, which lives naturally at high pressures, up to the ocean surface, the gases dissolved in their body fluids will expand and blow the guts out through the mouth. The fish will explode. But that's a fish.

The truth is that, while a pressure of zero pounds per square inch is very unhealthy for humans, it won't make us blow up. We will suffer violently from the bends, all our internal gases would rush out of our body orifices—calling it farting just doesn't cut it—and we'll lose consciousness in only a few seconds. The proof came in 1971, when a Soviet spacecraft underwent accidental depressurization, exposing the three cosmonauts within to the vacuum. Their bodies did not explode or become deformed with exposure to the vacuum. Unfortunately, they died from lack of air. Dr. Pickover estimates, "You should have fifteen seconds of useful consciousness before you pass out, and several minutes would be required before you die." Apes exposed to a vacuum suffered some bleeding from areas with blood vessels very close to the surface, such as nasal passages, eyes, and lungs, but survived a brief exposure without permanent damage. Since humans and apes, which evolved on Earth, can survive the lack of pressure in a vacuum, it's not unreasonable to imagine that a life-form could develop in the vacuum and survive there.

An organism is, in essence, a contained packet of liquid and chemicals, like a Ziploc bag filled with water. If the skin of the packet is strong enough, it can hold itself together. Since the slug appears to have no nasal passages or eyes, those possible sources of danger are eliminated.

Another huge threat to the slug remains, however. The slug's mouth and throat are also exposed to the vacuum. And if it doesn't have some sort of airlock device to seal off its throat, the slug's entire digestive system would be exposed to the vacuum. If the cells lining the digestive tract are as strong and protective as the outer cells of the slug, no problem. But the whole purpose of the digestive tract is to absorb nutrients. To do this, cell membranes must be permeable, meaning they must be able to pass materials in and out. If they are, then the water in the cells will quickly evaporate into space. Indeed, when Han and the others step out into the slug's throat, it's moist and misty, suggesting the slug is losing water. Dr. Jakosky points out that "Any water lost would have to be replaced, presumably by ingesting new water, and it's not obvious what the source would be."

Their excursion into the "cave" raises more problems. Any mist in the slug's open mouth should be quickly sucked out into space. And since the slug's mouth is open, the pressure inside the throat would be zero, which would disable Han, Leia, and Chewie in fifteen seconds, with all their bodily gases rushing out of them. That would make for an interesting scene.

If the slug had an airlock device of some kind in its throat, which would allow food to be passed safely from the vacuum outside to a pressurized interior, we might be able to explain some of this. But when Han flies the ship back out of the slug, there is no such barrier.

Scientists do believe that life may be able to survive in space, but only hardy bacteria, probably in their dormant spore form, buried deep within the rock of a large meteoroid. Dr. Pickover points out that terrestrial bacteria were found to have survived on a camera left in the near-vacuum and extreme temperatures on the Moon for three years, when they were retrieved by the crew of Apollo 12.

So might a more advanced life-form possibly live in the vacuum of space? Dr. Stephens says, "I don't think the thing is even remotely feasible." Yet Dr. Jakosky feels life may be more varied than we know. "I'd be reluctant to rule it out."

HOW MANY ALIENS DOES IT TAKE TO START A BAR FIGHT?

We discussed earlier how unlikely it would be for aliens from many different planets to survive in a single environment, such as the Mos Eisley cantina. Dr. Pickover raises an additional problem besides the environmental one. "The senses of aliens could be very different. Communication would be difficult." He points out that every Earthly species perceives the world differently. "They can smell what we cannot, they can see what we cannot, they can hear what we cannot." Bees, for example, can see ultraviolet light invisible to us, and dogs can hear sounds undetectable by us. "If the organisms of the Earth were somehow able to describe their world to you," Dr. Pickover says, "it would probably not be recognizable to you. It's likely that we will never be able to fully understand alien ideas, just as we may never be able to understand the 'language' of dolphins." Misunderstandings would be common unless one was in a very close relationship with an alien, like Han's friendship with Chewbacca.

Among humans from the same country, misunderstandings can arise easily enough. And among humans from different countries, language and culture can raise more problems. Considering that aliens might see, smell, and sense different things than we do, it's not surprising that an alien in the cantina might dislike Luke for seemingly no reason. It's a short step from there to violence, and before you know it, the barkeeper has more body parts to clean up.

WHEN THE TEDDY BEARS HAVE THEIR PICNIC

We discussed the Ewok moon in the last chapter, considering various elements of the environment, and found that it could potentially support life. Let's now look at the life that has evolved there.

The Ewoks are short, furry creatures that look a lot like teddy bears. They walk upright, have short limbs, and short fingers and toes bare of fur. Each hand has an opposable thumb, as on humans, allowing the Ewoks to use tools. Their fingernails and toenails look like humans' and are kept neatly trimmed. In *Return of the Jedi* we learn that they set traps for food, so they are predators. According to *The Star Wars Encyclopedia*, they are not only hunters but gatherers, which makes them omnivores. The Ewoks live in tribal groups, in villages built high up in great trees of the forest.

The Ewoks' small size would be a handicap in their development of tools and technology. It would allow them to wield a tool with only about one–twenty-fifth of the energy we could, making it hard to imagine how they might chop wood, build their homes, or kill an animal with one of those spears. Their stubby fingers would make fine work very difficult. And their short arms make it hard to imagine the Ewoks starting a fire without burning their noses. But perhaps they're particularly dexterous and patient.

The terrestrial animals that Ewoks most resemble are koalas, Australian natives about thirty inches tall. These short, furry creatures have rounded ears that perch on top of their heads like the Ewoks'. The color and pattern of koala fur varies with the individual, also like the Ewoks'. Koalas have long fingers, including two opposable thumbs on each hand and one on each foot. These allow the koala to hold branches in a powerful, viselike grip. Koalas also have long claws to help them climb trees, and rough pads on the underside of their hands and feet to increase their traction while climbing. Koalas live in loose groups, but they each like to have their own tree to live in. They are vegetarians, living on eucaplyptus leaves.

Comparison to another terrestrial species, chimpanzees, is also illuminating. Chimps live in groups of fifteen to eighty and build nests up in trees, as Ewoks do, sleeping in them at night to avoid predators like lions and tigers. While chimps are mainly vegetarians, they will sometimes kill baboons or pigs for food. Chimps climb trees using their powerful, grasping hands and feet, each of which has an opposable thumb. They also get around by swinging with their long arms from branch to branch.

We immediately see several key differences between Ewoks and these other species. Ewoks lack the characteristic traits of tree-dwelling vertebrates. Dr. White points out that such animals have at least one of two qualities: "One, a bunch of sharp claws on their hands and feet that they cling onto the tree with. Two, a quadrumanous or four-handed anatomy. If you look at the foot of a primate, it looks like the hand, and it can grasp." While Ewoks have opposable thumbs on each hand, their fingers are really too short and fat to use these thumbs to grasp any but the skinniest branch. And their feet have no opposable digit. It's hard to imagine them scaling the huge trees that surround them, since their arms and legs could not wrap around them, and their nails could not dig into the bark. Anthropologist Michael Seaman finds the Ewoks' body structure unlikely. "They don't look like they're really adapted for anything."

We could argue that Ewoks don't naturally live in those huge trees, just as humans don't naturally live in apartment buildings.

The Ewoks may have simply decided to move into the trees because it appealed to them, or because it offered them shelter from ground-dwelling predators. And they might use stairs and ladders to climb up there, rather than climbing up unaided. Michael Seaman finds this more plausible. "They may have evolved in a different type of forest or scurrying along the ground." In a forest of smaller trees, their stubby limbs could be more effective. They could put one foot on either side of the trunk and push their bodies up, like we climb a rope.

If they did evolve in a forest of small trees, though, what happened to it, or why did the Ewoks move to this forest of giants? Perhaps the Ewoks simply multiplied and spread beyond their original habitat. Perhaps they were displaced by Imperial forces. Or perhaps they found the great trees better homes for more extensive, elaborately constructed villages, where they could really put on a great Jedi barbecue.

THE SMALL, THE BIG, AND THE HAIRY

To consider the final group of aliens in the chapter, let's return to the planet Tatooine. In Chapter 1, we theorized that the planet may have gradually dried over billions of years. How would life adapt to the desert environment? Surviving under such harsh conditions requires that organisms cope with extremes of temperature, survive with little water, and make the most of what water they have. On Earth, a wide range of animals and plants have adapted to life in the desert. They have many mechanisms for dealing with the lack of water, some external, some internal. We have no information about what internal mechanisms might be employed by Tatooine dwellers, but we can deduce a fair amount about how well they're adapted to a desert environment by their external characteristics.

Small mammals tend to deal with the heat by retreating to underground burrows during the hottest part of the day. Only a few inches below the ground, the temperature can be significantly cooler. Since most of these creatures cannot pant or sweat to release excess heat, it's critical for them to be able to escape the worst of the heat through behavior. This solution to the heat is limited to small animals, since the larger the animal, the harder it is to dig an underground burrow of sufficient size.

On Earth, the Jerboa or desert rat stays in such burrows during the day, where the temperature usually reaches no more than 68 degrees. When they come out at night, they move so quickly that they look almost like a cartoon Roadrunner blur. Their long hind legs allow them to leap like a kangaroo up to 6½ feet, and their large feet, a bit like snowshoes, help them move quickly over loose, sandy soil. This jumping motion keeps the contact between their feet and the hot sand to a minimum. Another trait they share with the kangaroo is a long tail that helps them keep their balance, even when they suddenly change direction. These traits allow them to move quickly with minimal exertion, a very valuable ability to have in the desert, where you want to get food and get home before sunup.

The Jerboa are quite good at functioning on little water. They extract all their needed water from the seeds they eat. They're also much better at conserving water than humans are. Since they don't pant or sweat to lower their body temperature, they don't need as much water as we do. They excrete very little water, 20 percent less than a regular rodent. They also seal up their burrows during the day, locking in the moisture, and they sleep with their mouths lying next to stored seed, the moisture in their breath going into the seed to be eaten and recycled later. While different deserts are homes to different species of small rodents, they all have these elongated back legs and shortened front legs, suggesting that these are very valuable characteristics to have.

We see strikingly similar creatures in Mos Eisley in the special edition of *A New Hope*. They scatter as Luke and Obi-Wan drive into the city. These little Scurriers have short front legs, long hind legs, and long tails, and they usually run on their two hind legs, just like the Jerboa. While we do see them out during the day, this may be because their normal rhythms have been disrupted by man. Or perhaps daytime is the best time for scavenging on Tatooine, and a quick scurry from one air-conditioned building to another may keep them sufficiently cool.

Tatooine also has its share of larger species. Large animals take longer to heat up than their smaller counterparts, giving them more tolerance to heat. If you put a glass of iced tea and a gallon of iced tea out in the sun, you would expect the glass of tea to grow warm more quickly than the gallon. The smaller the object, the more quickly it will heat or cool. So once an animal is too large to burrow into the ground and must stay out in the sun, the larger it is, the better.

The Dewback is a reptile indigenous to Tatooine. We see stormtroopers riding this large, ungainly, four-legged creature. It has gray-green skin, a thick muscular tail, skinny legs, and bird-type feet, and performs a hitching waddle through the sand.

While reptiles cannot withstand extremely hot temperatures, they do have several advantages over other types of animals in a desert setting. Their tough, scaly skin keeps water loss to a minimum, and their eggs have leathery shells that keep them safe from drying out. But desert-dwelling reptiles have a hard time dealing with the heat because they are cold-blooded. Their temperature is not internally regulated, as ours is, but simply rises or falls depending on the environment. This makes controlling body temperature one of the top priorities of any reptile. Exposed to the sun on a hot desert day, most reptiles' temperatures will rise too high for them to survive. No reptile can survive a temperature of over 118 degrees. My iguana, Igmoe, basking outside on our deck on a sunny 85-degree New Hampshire day will begin to pant after an hour or so, struggling to release excess heat. One method reptiles have of controlling their body temperature is to manipulate how much of their body is exposed to the sun. To maximize heating, they orient their long, cylindrical bodies with one side toward the sun. To minimize heating, they aim either their heads or tails toward the sun, exposing as little surface area as possible. Even then, though, temperatures will often rise above the point they can tolerate. To cope with this, most reptiles are active in the morning and evening only, retreating to rock crevices or burrows at the hottest part of the day.

Hopefully the stormtroopers don't force the Dewbacks out into the sun at temperatures above what they can tolerate. The Dewbacks seem to cope reasonably well in the daytime situations where we see them. By manipulating their orientation to the sun, they can at least minimize the heat they absorb.

Another technique for coping with the heat is used by the agamid lizards. They have long legs that hold their bodies up off the burning sands. This offers a small bit of relief. In addition, the agamids, when stationary, always keep one foot in the air, constantly switching feet by turns in a circular cycle, so no one foot will get too hot.

The Dewback's skinny legs are tall enough to keep its underside a foot or two above the hot sands, so it can avoid that intense heat just like the agamids. I'd like to think it might also alternate feet, though we don't see a Dewback long enough to observe this.

The Dewback's feet are its most troubling aspect. They are quite birdlike, with two toes pointing generally forward in a V and one toe pointing straight back. While the spread of the toes will help distribute the weight of the Dewback over a large area, making this type of foot superior to a hoof, the narrow toes will not be terribly effective at pushing through loose sand. The camel, for example, has webbing between its two broad toes, and a foot as big as a plate. The arrangement of the Dewback's toes suggests it does not normally live in areas of loose sand. Most likely, it lives in more stony regions of the desert, as many reptiles do. There, its tough bird feet would serve it well. But on loose sand they could easily get bogged down. I wouldn't suggest stormtroopers ride them into the Dune Sea—or then again, maybe I would.

One final interesting characteristic of the Dewback is its name. As night falls on the desert, the ground cools rapidly, cooling the air directly above it as well. Since cool air can't hold as much moisture as warm air, the cool air deposits a thin layer of dew on the ground. What does that have to do with the Dewback? Well, one terrestrial reptile, the thorny devil, uses dew. Its skin cools rapidly as night falls, triggering dew to condense on its body just as it does on the ground. This dew then runs into hundreds of folds in the thorny devil's skin, the folds channeling the droplets of dew to the thorny devil's mouth, where it can drink them. The Dewback may have a similar mechanism, dew forming on its back and running through the wrinkles and folds in its skin to a place where the Dewback can access it.

Another reptile that lives on Tatooine is the Ronto, a long-necked beast of burden that looks like a new species of dinosaur. The Ronto has a number of characteristics that suggest it may be well adapted to desert conditions. The long neck of the Ronto is a characteristic it shares with a number of desert dwellers. Giraffe-necked antelopes not only have very long necks that allow them to eat leaves out of reach of other animals, but they can stand on their hind legs as well to reach vegetation even higher up. The Ronto may similarly use its neck to reach food, and we see it rear up on its hind legs. The camel also has a long neck, as we discussed earlier. The height of the camel's head—a result both of its long neck and its very long legs—elevates it above the level of most sandstorms. Sand particles tend to be of a rather uniform size. Thus when the wind blows at a certain velocity, the height that the sand can be lifted is fairly predictable. Depending on the ferocity of the storm, sand may rise from 3 to $6\frac{1}{2}$ feet off the ground. Below that height, the air is filled with stinging sand that can invade your nose and mouth. Visibility is reduced to almost nothing. Above that height, the air is almost completely clear. In a sandstorm, a camel's head rises above the level of the sand, allowing it to breathe and see with little problem. The Ronto would share this advantage.

The Ronto's legs raise its body higher than the Dewback's, so its stomach is about a Jawa's height off the ground. A hot layer of air tends to form in the few feet just above the desert surface, and the Ronto's legs would help to elevate it above this layer. In addition, the Ronto's large round feet resemble those of a camel. About the size of a serving plate, they would help keep the Ronto from sinking into the sand.

The Ronto's neck is thick and bony, and it has a bony bulge behind its head. Different species of dinosaurs had odd bony structures in the head and neck area that may have served various purposes. The protoceratops had a bony growth on the back of its skull that looks rather like the bulge behind the Ronto's head. The triceratops had a wide bony frill like a crown protruding from the back of its head.

By studying the chemical composition of this frill, scientists recently discovered that it helped the dinosaur radiate excess heat. Dr. Reese Barrick, from the University of North Carolina, tested various sections of the triceratops's skeleton for levels of different oxygen isotopes. An isotope is simply a different version of an element, with either more or less neutrons in its nucleus than the element usually has. Oxygen normally has eight neutrons in its nucleus. But sometimes a heavier oxygen is found with ten neutrons. The heavier oxygen tends toward cold locations, while the lighter tends toward warm. As the triceratops grew, the bones in colder parts of its body had more heavy oxygen incorporated into them, while bones in warmer parts of its body had more light oxygen incorporated into them. Thus by measuring the levels of the different oxygen isotopes in various bones, Dr. Barrick could deduce the temperatures at different points in the triceratops's body. What he found was that the frill was warm in the middle and cool on the outside, showing that it helped radiate heat away.

Similarly, desert jackrabbits use their large ears to radiate heat, as we discussed in connection with Jar Jar. The jackrabbits' ears have a dense network of blood vessels that bring the hot blood near the surface of the skin and allow it to release its heat to the environment more quickly. The cool blood then returns to the interior of the body, cooling the rabbit. Scientists discovered signs that the triceratops's frill similarly had a network of blood vessels crisscrossing it.

While scientists aren't as certain about the purpose of the protoceratops's bony bulge, one possible explanation for it—and for the Ronto's oddly shaped head and neck—is that the structure helps to radiate excess heat and keep the animal cool.

Another large species on Tatooine is the Bantha, an elephant-sized mount used by the Sand People. While Banthas are not native to Tatooine, they appear to function quite well in the desert heat. Their most striking characteristic is their long, thick fur. You might think that fur would be a horrible hindrance in the desert, overheating the Banthas like a fur coat. Yet fur insulates an animal, keeping out excess cold or heat. Many desert animals have fur, from the jerboa to the antelope to the camel. In the desert, the sun's heat is actually absorbed by the fur on the animal's back, preventing it from penetrating deeper into the skin. The hair on a camel's back has been measured at 158 degrees, while the body temperature of the camel was only 104. The Bantha's fur would certainly help it cope with the heat.

The Bantha also has a long, furry tail. While we see it dragging along on the ground, it may be capable of significant movement, as most animals' tails are. The Namib ground squirrel fluffs out its tail and holds it up over its head like a parasol, to shelter itself from the sun. Perhaps the Bantha does the same.

There's one more nonhumanoid Tatooine dweller we have to discuss. The Sarlacc, according to the *Star Wars Encyclopedia*, is not a native of Tatooine, but seems to function in the desert climate just fine, digging itself into the ground and waiting for prey to come. We can't be sure how big the Sarlacc is, but it must be fairly large to have such a huge appetite. While burrowing seems limited to smaller animals on Earth, the Sarlacc somehow manages to get its huge bulk into the ground. For a terrestrial model for this kind of behavior, we look to a much smaller animal, the ant lion.

Ant lions live in a variety of climates and are common in the southwest United States. In their larval stage, ant lions have a large head, spiny jaws, and a bristly body about $1/2$ inch long. Moving backwards, the larval ant lion traces out a circular pattern, spiraling steadily inward, digging deeper and deeper, until it creates a steep, conical pit in the sand and buries itself at the base of it. All that remains visible are long, curved jaws that lie open waiting for prey.

When an unlucky ant comes up to the edge of the pit, the sand collapses, and it falls down into the trap. The ant—much like Lando Calrissian—finds it can't climb out of the pit. The sides are angled so they crumble when the victim tries to crawl out, which is just what happens when Lando tries to climb out of the Sarlacc's pit. In the rare event that the prey looks like it might escape, the ant lion flicks sand at it, triggering an avalanche that brings the victim tumbling into its hungry maw. The ant lion snaps its jaws shut, injects a paralyzing poison and digestive acids into the victim, then sucks out its vital juices. When the ant lion is finished, it flings the carcass out of the pit with a flick of its head.

Although the body of the prey isn't digested inside the ant lion for one thousand years, as is said of the Sarlacc, any juices it has extracted from the prey do remain in the ant lion's body, since it has no method of excreting waste products. It's not until the ant lion transforms into its pupal stage—the inactive stage between larva and winged insect—that it can eliminate waste. This means that the ant lion must hold all its waste for its entire larval lifetime: three years. And I thought sitting through a movie could be tough.

DID YOU LEAVE YOUR HEADLIGHTS ON?

Tatooine has two indigenous humanoid species, the Jawas and the Sand People, also known as Tusken Raiders. We have never seen a Sand Person without his protective mask or wrappings, so it's hard to say whether they have specific characteristics favorable for desert living. Obviously they need artificial aids to survive in the harshest desert conditions, just as we need aids—coats, gloves, and boots—to survive in the winter. If a species is intelligent enough to use artificial means for survival, they don't need to be naturally as well adapted to the environment to survive. So aside from concluding that the Sand People are intelligent, we can't tell much more.

The Jawas also remain rather mysterious, swathed in robes from which their bright eyes pierce. The Jawas' hands appear furred, and the *Star Wars Encyclopedia* calls them "rodentlike," so we might conclude they are a furred mammal of some kind. They're hanging out among rocks and caves when they ambush Artoo. Perhaps they take shelter in caves during the heat of the day—or at least they may have lived this way before they had large air-conditioned transports for their comfort.

The Jawas' most striking feature is their glowing eyes. While those twin lights do at first impress us as eyes, though, it's not clear that this is truly what they are. They may be artificial, a tool, like a coal miner's light, to assist them in seeing into caves and other dark places. Yet if that is so, why do they keep them on during the day?

So perhaps we were right in our initial impression, and these two glowing disks actually are their eyes. Let's consider how eyes work. Eyes are sensitive light-reception devices. Light from the environment enters through the pupil, and that allows us to see our environment. If the eyes themselves glow like a flashlight, this intense outgoing light will interfere with the incoming light, in essence washing it out. Dr. Pickover offers an analogy. "How well could we hear if our ears emitted a continuous sound?"

Instead, the eyes and lights could be separate organs, one to receive light, the other to emit it. Dozens of terrestrial organisms emit light, including fireflies, earthworms, algae, fungi, jellyfish, crustaceans, and fish. Various chemical reactions can produce a bioluminescent glow. The most brilliant light from a single creature probably comes from the Caribbean fire beetle, which has a heart-shaped orange light on its stomach and two yellow-green lights on its shoulders. Women even put fire beetles in their hair as decorations—an idea, perhaps, for George Lucas, in his quest for unusual female hairstyles. The greatest collective light display is put on by male fireflies in Thailand, who gather in rows of trees and put on an impressive show of synchronized flashing to attract females.

The most useful terrestrial animals with which to compare the Jawas are deep-sea fish. Two-thirds of all deep-sea fish are bioluminescent. In the dark depths where a deep-sea fish lives, the lights on its own body may help it attract prey or mates, while the lights on other fish help it spot potential mates, recognize predators, and keep close to its school. These fish have light-producing glands called photophores. The photophores are composed of many tiny tubules, in which light-producing bacteria are confined. The bacteria and the fish live in a symbiotic relationship, the fish providing the bacteria a tasty enzyme to eat, the bacteria, in the chemical reaction that occurs when they eat the enzyme, producing light for the fish.

Two fish, Anomalops and Photoblepharon, known as lamplight fish, have photophores beneath each eye. These photophores can even be rotated in and out of a bony socket on the fish's face, like headlights on a fancy car, so the fish can make the lights "blink" by moving them in and out, or the fish can hide by tucking the lights inside. Muscles can even aim the photophores a bit more ahead, so they can function more like headlights. At chow time, the fish gather with their school and their lights illuminate the immediate surroundings, allowing fish to see the plankton they feed on. Looking at these fish head-on, the photophores look like two glowing eyes. If the Jawas did evolve in caves, their lights could serve the same function.

Even in environments that aren't completely dark, organisms find luminescence an advantage. One of the main uses of luminescence is to help attract the opposite sex. The female annelid fireworm releases streams of glowing eggs in the ocean. The males are attracted to the eggs, flash a light in response, and release their sperm. If the Jawas' lights serve this purpose, they all seem to be constantly looking for love.

Bioluminescence can also serve more devious purposes. The females of one species of firefly mimic the flashing pattern of another. When the males show up ready for action, the females, not interested in mating at all, gobble them up. Similarly, the Jawas might use their lights to lure prey close.

The ponyfish has glowing cells along the underside of its body. When predators below look up at the ponyfish, the glowing cells help the fish blend in with the light-mottled surface of the water above. Perhaps bioluminescent fungi grow inside the caves, and the Jawas' headlights help them blend in with the cave wall, camouflaging them from predators.

The lights might be particularly useful in the desert. Before, we talked about how the camel's height—and the Ronto's—lifts its head above the level of sandstorms, allowing it to see and breathe. Jawas, as short as they are, will be completely immersed in a sandstorm. Many short animals can move very quickly to get to shelter in times of need. The Jawas don't seem particularly fleet of foot, though. If the Jawas' lifestyle requires they travel significant distances from home to find food, they could be caught out in a sandstorm. In such a case, their glowing eyes could serve as beacons, helping the Jawas find each other.

So what good is it to find your friends in a sandstorm if you're all lost? The camel may shed some light on this. Camels out in the heat of day huddle together. This seems odd; we might expect them to stand separately. But remember that a small animal will gain or lose temperature more quickly than a large one. When the camels group together, they are making themselves, in essence, a single, larger organism. So their temperature is affected less by the environment. Similarly, Jawas in the extreme heat and dehydrating winds of a sandstorm might want to find each other so they can huddle and create a larger collective organism, perhaps helping them survive until the air clears.

AT HOME IN THE JUNDLAND WASTES

Finally, let's consider the human settlers who live on Tatooine. How do humans cope in desert environments? Tatooine residents seem to favor thick hooded cloaks and ponchos. Obi-Wan, Qui-Gon, Luke, the Sand People, and even the Jawas seem to prefer this type of dress. You might think that you'd want to wear as little clothing as possible in the desert heat (or just enough to avoid a sunburn). But heavy robes provide the same insulation that the fur of animals does, helping to moderate the temperature beneath from extremes of the environment. Bedouin Arabs wear thick robes and head wrappings for the same reason. Loose clothing that traps a layer of air beside the body is the best.

Many people believe white clothing is much better to wear in the hot sun, since white is better at reflecting heat than black. This is only partly true. White does reflect visible light, while black absorbs it. Yet most of the heat in the desert comes not from visible light but from lower frequency infrared light. Infrared light is absorbed as well by white clothing as it is by black. Thus the brown cloak of Obi-Wan and the brown cloaks of the Jawas are perfectly suitable to the desert.

Even with such clothing, though, humans must constantly struggle to survive in the desert. During sandstorms, the temperature rises and the air becomes very dry. A person can lose up to a quart of moisture from his body in an hour. Without lots of drinking water, a person can die within a few hours, literally drying into a mummy.

And conditions aren't much better even when it's not a sandstorm. Humans cool themselves by sweating, which quickly depletes their bodies of water. If Luke Skywalker were abandoned out on the desert at twin-sunrise with no protective clothing or shelter, he would sweat away up to twenty-one pints of water before nightfall. His body would draw water from his fat, tissues, and eventually his blood. As his blood thickened, his body temperature would rise, as if he had a fever. Blood circulation helps to cool blood, by bringing it just underneath the skin where it can radiate heat away. Thus impaired blood circulation makes matters even worse. He would not survive a single day. Uwe George tells the story of a couple visiting the Sahara. They decided to drive their car from a large oasis to a small oasis twenty miles away. They arrived safely at the small oasis, and after a visit turned around to drive back to the large oasis. They assumed their return trip would be as uneventful as the initial one had been, so they didn't bother to fill their water bottles. They also forgot to fill their gas tank. They ran out of gas ten miles from the large oasis. The woman decided to wait in the shade beside the car while the man went ahead to the large oasis for gas. When the man returned five hours later with the gas, Uwe George says, "She was still sitting there. But she had perished of thirst." Hopefully Luke keeps his speeder stocked with water at all times, in case of such a situation.

The desert is not friendly to machines, either. Sand mires cars in dunes and chokes up car engines. While we're on the subject of the speeder, I can't imagine why one would use an open vehicle to travel in the desert. Loose sand would irritate ones eyes and nose, and a sandstorm would cause major breathing problems.

How could one live comfortably in such a place? The home of Owen and Beru Lars, where Luke lives, is built into the ground. Desert dwellers on Earth have used a similar strategy to create cool homes. In the Tunisian village of Matmata, over one hundred homes have been tunneled into the ground, each with a central courtyard open to the sky. Such homes are usually two stories deep, the upper story used for storage, the cooler lower story used for living. Just as animals dig burrows to keep cool, humans do the same. When you look down from ground level into the circular courtyard, you see numerous doors and windows in the walls, and stairs connecting one level to the other. The scenes in Owen and Beru's home were actually filmed inside such a structure, the hotel Sidi Driss in Matmata.

Having a courtyard about thirty feet below ground level means it will more often be in shade, receiving the direct heat of the sun only when it is nearly overhead. The courtyard also tends to retain cool night air and keep the inside of the house cooler during the day. Thick walls further insulate the inner rooms, creating a home significantly cooler than one built above ground. We see those thick walls again in Anakin's home in Mos Espa, helping to insulate the interior even though it's not below ground. Apparently, air conditioning is out of fashion on Tatooine.

Star Wars presents a universe filled with an amazing variety of life, filling every available ecological niche. While scientists remain uncertain about what sort of alien life we will find, it is sure to include species as bizarre as those we see in the movies, and probably species even more bizarre. But visiting "a galaxy far, far away" may give us some small hint about what may be waiting for us out in our own galaxy.

Aliens aren't the only strange creatures we meet in *Star Wars*, though. In the next chapter, we'll discuss an entirely different class of creatures that are not aliens at all, but artificial life-forms: droids.

3

DROIDS

Why I should stick my neck out for you is quite beyond my capacity.

—C-3PO, *A New Hope*

Tall one is almost always worried and unhappy, believing disaster around every corner. The short one is adventurous and determined, impatient with his cowardly partner. Anxious and frustrated, the tall one at times lashes out, verbally and even physically, kicking the short one or slapping him on the head. The short one counters by giving the tall one the raspberry and even, when their differences become too great, abandoning him.

No, this is not a dysfunctional couple in family therapy; it is a pair of robots: C-3PO and R2-D2. When most scientists think of a robot, they think of a sophisticated artificial intelligence, sensors that can detect the surrounding environment, some sort of limbs for interacting with the environment, and a method of travelling through that environment. Yet a funny thing happened on the trip through George Lucas's imagination. In addition to these characteristics, robots gained personalities, desires, and emotions. Droids can be kind, cruel, loyal, afraid, disgusted, excited, concerned, impatient, embarrassed, and proud. These characteristics cause us to become as emotionally attached to the droids in *Star Wars* as we do to the humans.

We've met many robots in science fiction movies, but none as memorable as R2-D2 and C-3PO. First of all, they are amazing embodiments of advanced technology, able to perform a wide variety of tasks. Artoo has sensors that can detect distant signs of life on Tatooine as well as details of his nearby environment, a hologram projector, and a very expressive nonverbal voice. He functions as a component of Luke's X-wing fighter; plugs into the Imperial network on the Death Star, reads files and overrides controls; carries a huge amount of information; and thinks creatively to complete his missions. Threepio doesn't seem quite as brainy, or at least doesn't like to admit he can follow Artoo's technical talk. He is conversant in six million forms of communication, speaking human and alien languages and communicating with machines, including the *Falcon*. Threepio calls Artoo his counterpart, which suggests they work together in some way.

But it's not what these droids do that makes them so memorable; it's who they are. Artoo and Threepio have personalities as strong as the humans in the *Star Wars* universe. In fact, they seem so much like living beings that I find myself unable to call either of them *it* as I write about them. Both Artoo and Threepio have strong ideas, goals, and emotions, and they bicker endlessly, though affectionately, like old friends. We sympathize with them, just as we do with the human and alien characters, as they go through their adventures.

Might robots like Artoo and Threepio someday be within our reach? And if they are, are these the type of robots we'll be likely to create? Before we talk about robot personalities, intelligence, or emotions, let's take a look at the basic shape of these future robots.

WHEELS OR LEGS?

Artoo normally cruises along on three small tread wheels like a tricycle, which is fine for the polished corridors of the Death Star. When he gets onto rougher terrain, like the swamps of Dagobah or the stairs leading into the Mos Eisley cantina, he has to bring his two back wheels up alongside his body and use them as stubby legs. From watching Artoo, one can tell that these stubby legs don't work terribly well. They're very short, and lack the joints that help human and animal legs function. Obviously Artoo has been designed to function in a mainly indoor, sophisticated environment, or else to remain stationary, as he is while plugged into an X-wing.

Most robots today are stationary, designed for a specific purpose, and set up in the location where they can accomplish that purpose, such as on an assembly line. Those that require mobility are usually built with wheels. A number of hospitals now use wheeled robots to deliver lab specimens, surgical supplies, medical records, and meals. Wheeled robots are simple to make, energy efficient, and can easily navigate interior environments where floors are smooth. Even some robots designed for outdoor use have been made with wheels, such as *Sojourner*, the wheeled rover NASA recently sent to Mars. While *Sojourner*'s mission was a success, its mobility on the rocky surface of Mars was limited.

For movement on uneven, outdoor surfaces, you might think we'd want to create two-legged humanoid robots such as Threepio. After all, we have two legs, and we do a pretty good job of getting around. Yet of the robots currently in outdoor use to clean up hazardous-waste sites, dispose of bombs, or put out fires, none are bipedal. Why? A two-legged gait requires a lot of coordination.

As you walk on two legs, your body must shift slightly from side to side so that your center of gravity is more nearly over the supporting leg. Sixty percent of human bodyweight is above the hips, so balancing this top-heavy structure is not easy. You must also learn how to use the muscles in your hips, thighs, knees, calves, and ankles to move one leg forward and hold the other stiff. Walking uses more than thirty muscles in each leg and yet more muscles in the trunk. So it actually is quite a challenge to walk and chew gum at the same time.

In designing robots, scientists find it difficult even to artificially duplicate a bipedal stance. Simply standing still and balancing your body isn't as easy as you might think. Stand up right now and notice what's going on in your body (this is the exercise portion of the book). Your leg muscles will make tiny adjustments as your torso moves slightly forward or back; in fact, you'll find it impossible to completely relax your leg muscles and remain upright. (You can sit down now.) The two-legged stance is inherently unstable.

While four legs provide a much more stable stance—ask any table—walking on four legs still requires a shift from side to side in the center of gravity. Five legs are the minimum allowing a stable gait. A five-legged robot can have one leg in the air and still remain stable without shifting its center of gravity. This makes the coordination issues much less complex and the gait much easier to artificially duplicate. Of course, if a robot has five legs and can only lift one at a time, it's not going to be terribly speedy. Scientists have found that modeling robots after insects, with six legs, provides a stable stance and gait, and allows the robots to lift three legs at a time.

Insect-modeled robots can use a tripod gait, the front and back leg on one side hitting the ground at the same time as the middle leg on the opposite side. With three legs on the ground at any time, both insects and the robots based on them are extremely stable. Dr. Randall Beer and colleagues at Case Western Reserve University have built several such robots around 1½ feet long. Although their first robots were slow and awkward, the latest models are graceful and almost lifelike.

By studying the movements of the cockroach and the neurons triggering those movements, the team has learned that a combination of centralized and decentralized signals control the cockroach's legs. Centralized control coordinates the movements of the legs and keeps the cockroach balanced, while decentralized control allows each leg to act more independently. The closer the team makes its robots to actual cockroaches, the better the robots have performed.

These robots are still not nearly as fast as real cockroaches, though. We have a long way to go before we match their agility and speed. Yet robots modeled after insects can skillfully navigate uneven terrain. They've been used to explore an active volcano in Alaska and clean up nuclear power plants. They are being developed for a number of future uses, including the exploration of other planets.

Must all legged robots have six legs, then? Well, founder of the MIT Leg Laboratory Dr. Marc Raibert and colleagues built a robot with only one leg, the most unstable situation there is. Working with just one leg allowed Dr. Raibert to focus on issues of balance, rather than worrying about how to coordinate the movement of different legs. The robot maintains its balance by continuously hopping, just as a person on a pogo stick stays upright by hopping. It can hop in a particular direction or follow a particular path. Its top speed is a little less than 5 miles per hour. While scientists don't particularly believe a one-legged robot is the wave of the future, they do believe that a two-legged robot that runs as a human runs, with just one leg hitting the ground at any time, is essentially the same as a one-legged robot.

Using this one-legged success, a two-legged robot has indeed been built. With its two legs hopping in turn, it can maintain its forward/backward balance, propel itself ahead at up to 13 miles per hour, and even bound up stairs. Unfortunately, the robot cannot maintain its lateral, or side-to-side balance. A rotating boom extending from the center of the lab to the top of the robot keeps it laterally stable, constraining it to hopping around in a circle.

Robotocists at Honda decided to take a different approach when they began their humanoid robot research in 1986. Their goal was to create a robot that could operate in people's homes and interact with them. Rather than trying to create a biped that ran, they thought maybe they should first teach it to walk. With detailed studies of the joints in the legs and feet, they built robot legs that would have the same range of motion and ability. They also explored the many ways that humans sense their state of balance, velocity, and direction, and created similar sensors for their robot. They have now built three prototypes, each progressively smaller, lighter, and more skilled at walking. The latest, P3, is about 5½ feet tall and 300 pounds. It can walk up and down stairs and transport objects in its arms. Yet it takes slow, deliberate steps, its speed limited to just 1 mile per hour. The robot has a blocky, bulky appearance, as if it's made out of giant Legos, with wide legs and feet that help it maintain balance. While it has some ability to maintain its stability on uneven terrain, Honda scientists are still working to improve this ability.

Although bipedal robots are proving a challenge to build, they are appealing because they could be built to resemble humans, rather than cockroaches or pogo sticks, and so would be easier for us to interact with. C-3PO is obviously designed to resemble a human, which makes his job of translating speech and following the customs of various humanoid races easier. A pogo stick would have a hard time bowing, shaking hands, or communicating through body language and gestures.

While Threepio's form may put humans at ease and foster communication with them, it is not terribly mobile. Threepio is a bit better able than Artoo to navigate stairs and uneven terrain, yet his range of motion is more restricted than a human's. He seems barely able to navigate stairs, let alone rocky or swampy terrain, and out on the sands of Tatooine his joints begin to freeze up. Obviously modeling Threepio after a man has its disadvantages. Do the advantages outweigh the disadvantages, making this the best design for Threepio?

Threepio's knowledge of etiquette and protocol suggests he is meant to function among ambassadors and diplomats. In such a rarefied atmosphere, in elegant surroundings, we can imagine that Threepio could get around fairly well, and his golden, humanoid appearance would add a touch of class to the proceedings.

But translating and assisting the elite aren't Threepio's only duties, much as he might wish they were. His stated specialty, "human/cyborg relations," is a bit confusing, since cyborgs are organisms with mechanical or electronic components, and the only cyborgs we see in the movies are Darth Vader, Luke after he receives his bionic hand, and Lando's assistant, Lobot. I think perhaps what Threepio means is human/machine relations, since he can serve as interpreter to the *Falcon*, to moisture vaporators or load lifters, telling their owners what is wrong with them. This is the capacity in which Luke's Uncle Owen wants Threepio to work, serving as a diagnostic tool for machines and other droids. Threepio seems to have an unnecessarily sophisticated design for such a purpose.

We have similar devices today, such as computers that diagnose problems in the transmissions of our cars. These computers are immobile. We simply bring the machines to be repaired to them. If the machines are too big to bring in for repair, the diagnostic computer could be mounted on a simple transport device. Uncle Owen could put it on his speeder or use one of the antigravity devices that seem fairly common. Perhaps a simpler droid like that is what Owen is looking for, and why he's hesitant to buy a fancy droid like Threepio.

So while Threepio's design makes sense if his primary function is indeed translating for the elite and making sure etiquette and protocol are followed at official functions, he would be a rare droid indeed in the galaxy, one of only a handful designed and built for such a specialized purpose. Other droids specializing in human/machine relations and translation would likely be of a much simpler and cheaper design.

Droids with other purposes would seem even less likely to be bipedal. Battle droids, for example, designed as killing machines, would probably be built in a number of different configurations, depending on how they were to be used. The only situation in which a humanoid form would really be useful is one in which robots are taking the place of human soldiers; they could operate the same equipment. If human soldiers are not used in specific combat situations, droids would have little reason for looking human. For inner-city guerilla fighting, we could use a spherical droid held aloft with antigravity, similar in appearance to the baseball-sized remote Luke trains with in *A New Hope*, though perhaps bigger to allow for increased laser power and intelligence. If antigravity is not an option, then insect-modeled robots would be very fast and flexible. Such battle droids could inspire terror in the enemy. They could even have laser-gun turrets on their backs. If we're looking for an inexpensive option, an R2 unit could serve in a number of situations. Plug it into a STAP or other antigravity vehicle and you have a formidable weapon. For space battles, it could be inserted into an X-wing or other fighter and fly it. It could even operate a Walker. If computers aren't terribly expensive to build, then you don't even need an R2 unit to plug into various accessories. Just build an intelligent STAP. It doesn't need a humanoid robot to hold onto it and fire its weapons.

If we were part of the rebellion, and we wanted a droid to help us execute our secret missions, what type of droid would be the most useful? Princess Leia most likely had to make do with what was available. What if she'd had her choice of designs? Well, a droid that fits in with its surroundings would have a better chance of evading detection. But since little attention seems to be paid to droids, this seems a minor concern. One that can easily navigate rough terrain would be able to avoid capture and reach remote rebel bases.

An antigravity droid or insect-modeled robot, then, would

seem to be the most practical solutions. But can you imagine Threepio's head mounted on the hood of a speeder, or Artoo with the six legs of a roachbot?

THESE TANKS ARE MADE FOR WALKING

One of the most memorable battle scenes in *Star Wars* takes place on the planet Hoth, where rebels attempt to defend themselves against massive walking tanks, called All-Terrain Armored Transports, or Walkers. Walkers seem to have two potential advantages: height to see and shoot long distances, and the ability to cross rough terrain. These two qualities, while both valuable, actually work against each other. The top-heavy design that gives the Walkers their height advantage would also tend to destabilize them on uneven terrain. We already know that four-legged locomotion, like that of the Walkers, is unstable and requires that the "body" of the Walker shift back and forth to compensate. The massiveness of that "body" would make this a challenge. A lower-bodied, six-legged robot seems as if it would function better on rugged terrain. But if the Empire does have the ability to coordinate a shift in the top-heavy Walker's center of gravity with its gait, and if an antigravity tank is for some reason impossible, would this be a useful shape for a battle tank?

Using the same techniques that created one- and two-legged running robots, Dr. Raibert has created quadruped robots that run with different gaits, trotting, pacing, and bounding. Such robots are not yet laterally stable, though. And they'd provide quite a rough ride for the Imperial troops inside.

Robotocist Kimura Hiroshi and colleagues built a walking quadruped robot that resembles a miniature Walker. Its four sturdy legs move quickly, giving the impression of two people scurrying along, one behind the other. It, too, is not laterally stable. It moves forward while attached to a pipe running above it. Yet it can handle moderately uneven terrain, stumbling over obstacles but quickly regaining its forward/backward balance.

The most fascinating quadruped design is actually the oldest one, dating back to 1968. Ralph Mosher at General Electric built a four-legged walking truck. Eleven feet tall, it looked more or less like a truck taken off its wheels and placed on four tall legs. Rather than being run by computer, this one's movement was controlled by a human driver in the cab of the "truck." The truck had four controls, two connected to the driver's hands and two to his feet. As the driver directed one of the truck's legs down against the ground, as to take a step, the control would push against the driver's hand, reflecting the pressure of the ground against the truck's "foot." This gave the driver the illusion that the truck limbs were his limbs, and made control easier. The truck was able to move with agility, even climbing a stack of railroad ties. The drawback of the design, though, was that it had to keep three legs on the ground at all times for stability, and that it could move a leg only a small amount at a time, so that its center of mass remained above the three supporting legs. Because of this, its speed never got much above 5 miles per hour.

The Walkers also only lift one leg at a time and take fairly small steps. The *Star Wars Encyclopedia* tells us that Walkers can travel up to 40 miles per hour. I'm not so sure.

The oddest idea for a tank was patented in 1942. The inventor was sure it would catch on, since it would be nearly impossible for an enemy to target and destroy. It was a one-legged hopping tank.

I, DROID

Now that we've explored the shapes of the droids, let's look at their intelligence. Unlike current robots or computers, they have the ability to plan and perform extremely complex tasks independently. Artoo demonstrates this skill in *A New Hope* when Princess Leia gives him a secret mission to complete: to deliver himself and Leia's message to Obi-Wan Kenobi.

Artoo enters an escape pod and sees himself and Threepio safely to Tatooine. Once there, he seeks out a populated area, and after being bought by Luke's uncle, quickly gathers the information he needs. He lies to Luke, claiming to be owned by Obi-Wan, and so discovers the Jedi's location. Then, after discerning Luke's interest in Leia's message, claims the restraining bolt prevents him from playing the entire message, so Luke will remove the bolt. Without the bolt, Artoo is free to pursue his mission. He leaves Luke's home and heads off across the desert toward Ben's home.

This sequence of actions requires an advanced intelligence, an understanding of human nature, the ability to reason, and flexible planning and decision-making abilities. Such qualities are key elements of an artificial intelligence, or AI, a computer that exhibits the qualities of human intelligence. Scientists have been working to create an AI for forty years, and although they have not yet achieved this goal, they have made substantial progress with several different approaches.

The rule-based approach focuses on making a computer an expert in a particular, narrow subject area. By choosing a narrow specialty, computer scientists are able to program extensive knowledge about the subject into the computer. IBM's Deep Blue, which beat world chess champion Garry Kasparov, is one such system. The computer contains all the rules pertaining to chess and has the ability to weigh various possible moves and their outcomes to choose the best one. Such systems, however, quickly break down when the computer is presented with a problem outside its area of expertise. It can't extrapolate from the knowledge it's been given or make comparisons; it can only follow the rules. 2-1B, the medical droid that treats Luke after he's attacked by the Wampa ice creature on Hoth, may be an example of such a system. It has expertise in a narrow area, but appears to have minimal interpersonal skills—like many doctors.

Both Artoo and Threepio appear to have certain areas in which they are "experts," such as translation and piloting an X-wing, but they also exhibit a great flexibility to function usefully in a wide variety of circumstances. So they can't be exclusively rule-based systems.

A second approach scientists have taken is to create an AI that uses case-based reasoning. Rather than blindly following rules, the computer draws analogies, comparing the situation confronting it with other situations it knows, determining which are most similar, and drawing information from the comparisons. For example, Artoo probably knows where human settlements are on Alderaan and a number of other planets. To find human settlements on Tatooine, he could compare the geographic features of those other settlements with the geographic features he observed as the escape pod fell toward Tatooine, locating the most likely areas for human habitation. In fact, this seems to be what he does, suggesting that he has some case-based abilities. Such systems have had some success, though unless the comparison is straightforward, the case-based systems have difficulty determining which comparisons are suitable and which are not.

To do that, computers need to have some basic knowledge about the world, what we call common sense. Our common sense is built from things we've learned throughout our lives. While we tend to take it for granted, this knowledge is very difficult to impart to a computer. Both Artoo and Threepio exhibit a great deal of common sense. For example, they know that if Luke is trapped in the garbage masher, he will not suddenly appear beside the *Millennium Falcon*. They know that if the two walls of the garbage masher come together, they will squash and kill Luke. And they know that humans prefer to be living than dead, so that Luke will be happy if they are able to stop the garbage masher. All these things may seem obvious to us, but they are not obvious to a computer, unless this information is input into it.

Dr. Douglas Lenat has been programming common sense into a computer called CYC, short for encyclopedia. His goal is to give CYC one hundred million pieces of common sense. This knowledge will help it draw valid comparisons and make decisions more efficiently, eliminating impractical or undesirable solutions. Thus far, CYC has conducted much more effective searches for information than standard Internet search engines. For example, when asked to deliver a photo of "a strong and adventurous person," CYC delivered a photo with the caption, "a man climbing a rock face." CYC recognized that rock climbing is adventurous and requires strength.

A third approach to creating an AI is to build a system that can learn from experience. If we can give a computer the ability to perceive events around it and learn from those events, then potentially the computer can develop intelligence just as a baby does. To create such learning systems, scientists build neural networks, systems designed to mimic, in a crude way, the structure of the human brain.

Regular computers are governed by a single complex central processor. Yet the brain has no central control. The brain consists of roughly one hundred billion nerve cells, or neurons, with each neuron connected to ten thousand others. In imitation of the brain, neural networks consist of many simple processors without any centralized governing program. These simple processors are connected to each other similarly to the way neurons are connected to each other in the brain. Scientists believe the massive number of connections between neurons gives the brain its ability to process one thousand trillion pieces of information simultaneously. This huge degree of interconnectedness, called parallelism, allows many different signals to travel from one place to another at the same time. Steve Grand, Chief Technology Officer of Cyberlife Technology and Director of the Cyberlife Institute, which is devoted to the creation of advanced intelligent artificial life-forms, believes parallelism is key to the development of artificial intelligence. "Brains are really machines in which many things happen simultaneously. Only such massively parallel systems are capable of being intelligent." So far, neural networks are nowhere near as complex or extensive as the brain—they have only one-fiftieth the brain power of a cockroach—yet in a basic way they do reproduce the decentralized structure of the brain.

A neural network works like this. Computer scientists input a specific stimulus on the input side of the network. The stimulus is transmitted through these connections to other processors, and a signal is emitted on the output side of the network. By "training" the network, scientists can make it emit the desired output. But how do we train a neural network?

The connections are the key to the network, rather than the individual processors. These connections can be strengthened or weakened, and the process of changing their strengths reproduces, in a rudimentary way, the learning process in the brain. Neurons in the brain remember previous signals that have passed through them, and which other neurons those signals have come from. Based on this previous experience, neurons give more weight to signals from other specific neurons. Those connections are strengthened, while others are weakened. This process is going on, for example, when you are learning to play the piano or multiply and divide. Pathways are being established in the brain, making these tasks easier with practice. Scientists train the network by adjusting the strength of its connections, creating the appropriate pathways that produce the desired result. These adjustments mimic a basic level of learning, yet the network uses no logic at all.

Even though neural networks aren't yet terribly sophisticated,

they have the ability to master processes that aren't easily programmed. For example, they can recognize complex patterns, a
skill called pattern matching. Neural nets are now being used to
recognize patterns and predict trends in the stock market. Scientists are also testing them as components in electronic eyes. In the
future, they might help computers recognize people's faces. Right
now a computer can recognize a person's face only when it displays a constant neutral expression and is shown in a full-face,
straight-on shot. Strong pattern-matching abilities could allow a
neural net to recognize Leia's face under various lighting conditions, at an angle or partially obscured, when she is smiling or
frowning, or even with her hair in various exotic configurations.
Artoo and Threepio certainly have this ability.

Steve Grand believes connectionist techniques used in neural
networks are the most promising for creating artificial intelligence.
"Rule-based systems and case-based systems have had fifty years
to prove themselves and haven't exactly lived up to their prom-
ise." Grand judges that Artoo uses neural networks. "Intelligence
of the kind R2-D2 shows is by nature an emergent phenomenon,"
he says, meaning that it arises as a property of a group rather than
of any one member of that group. Grand gives an example. "You,
as a distinct and unified human being, are an emergent consequence of the interactions of all the billions of individual cells that
make up your body. There is no single cell in which 'you' reside."
Similarly, intelligence must be an emergent consequence of the
interactions of the many processors connected in parallel that
make up the neural net. Intelligence cannot exist in any single
processor. Yet to create intelligence through many interconnected
processors, we need to make them interact the way neurons do in
the brain. "The big snag," Grand explains, is that "we have almost
no idea how the brain works at all!"

Since Artoo and Threepio have pattern-matching abilities, we
can assume neural nets make up at least part of their systems. In
addition, the *Star Wars Encyclopedia* says Artoo and Threepio
have avoided the regular memory wipes imposed on droids, which
has allowed them to learn from experience. The ability to learn
again implies the involvement of neural nets.

Our best chance at creating a true artificial intelligence may be
in combining these three approaches. And the droids seem to do

just that. The most brilliant intelligence, though, will not make a good robot unless it can sense its surroundings, gathering information, learning, and interacting. Artoo and Threepio can see, hear, and even feel. How far away are we from creating robots with these abilities?

DO YOU HEAR WHAT I HEAR?

Threepio and Artoo have a variety of ways to sense their world. They both seem to see, hear, and have a sense of touch. Both Threepio and Artoo react when they are touched: in *The Empire Strikes Back*, Threepio turns when Han touches his shoulder; and in *A New Hope*, Artoo bows when Threepio touches the back of his head. Creating robots with any of these senses poses a fascinating and difficult problem.

We touched briefly on the difficulty of teaching a robot to recognize a person's face from different angles and with different expressions. That's just one small example of the difficulty of making robots that "see."

Certainly we can connect a video camera to a computer. Computers can store video input and manipulate it. But teaching them to actually "see" the images the video contains is a completely different issue. The computer must separate objects from the background and be able to recognize these objects and their significance.

Simply distinguishing where one object ends and another begins is difficult. Often the only clues are a change in color or texture. Objects that move create even more problems. In the company of people, the computer must not only recognize faces under different circumstances, but must recognize bodies in many different positions, carrying out a variety of actions. And it must recognize the significance of those actions. For example, Threepio needs to know that when Han raises his index finger straight up in front of him, as he does in *The Empire Strikes Back*, it means "Wait a minute," or possibly "Threepio, shut up."

Computers have been programmed to recognize a variety of two-dimensional shapes, such as letters of the alphabet or photographs of human faces displaying neutral expressions. It's more difficult to teach computers to recognize three-dimensional objects, since they might see them from a variety of angles. Some factory robots have basic three-dimensional recognition abilities. For example, they can tell that an object with four legs, a flat horizontal surface on top of those legs, and a flat vertical surface protruding above that, is probably a chair. Recently, two German scientists, Dr. Ernst Dickmanns and Dr. Volker Graefe, built a computer-controlled car. The computer can "see" moving objects well enough to drive on roads with or without lane markings, avoid other traffic, and travel at up to sixty miles per hour. Still, its ability to recognize objects is limited to a certain area of specialty. It will be some time before we have robots as visually capable as humans, or as Artoo or Threepio.

As hard as it is to create a robot that can see, it's equally hard to create one that can hear. Hearing plays a critical role in the functioning of both Threepio and Artoo. They must understand the orders of their human owners. But can a computer understand our speech?

This is actually a much more complicated task than you might think, because human speech is not at all uniform. While Artoo might be programmed to enunciate his whistles and clicks clearly and distinctly, and to always use proper robot grammar, humans aren't so easily programmed.

The problems associated with teaching computers to hear fall into two general categories: sound and meaning. First, the computer must accurately identify the sounds being made. Not an easy thing, since English has over ten thousand possible syllables. Often we don't even pronounce every syllable. And of those syllables we do pronounce, we don't each generate the same exact sound. Different people pronounce words differently. Sometimes even the same person pronounces the same word differently. Princess Leia seems to like using a British accent in some situations and an American accent in others.

Once the syllables have been identified, the computer has to separate them into words. If each word were spoken separately and distinctly, this wouldn't be too difficult. But we tend to run our words and sentences together. In fact, some *Star Wars* dialogue approaches light speed, with no pauses to indicate punctuation. My nominee for fastest line comes in *A New Hope* from Luke, who delivers the entire sentence as a single word, without a breath: "You know with his howling and your blasting everything in sight it's a wonder the whole station doesn't know we're here."

If the computer is able to successfully separate the syllables into the correct words (perhaps Artoo and Threepio could consult the script for assistance), it still needs to understand the meaning behind the words. Human beings often misunderstand each other. How can robots do any better? This task is far from straightforward. Each person expresses himself slightly differently; the meaning of a word often depends on its context; and a person's tone can completely change his meaning, for example changing a sentence from a statement to a question. Artoo and Threepio need to deal with all of these issues and to have sophisticated artificial intelligences in order to understand their human companions.

At this point, computers remain extremely limited in any *understanding* of speech. They have, however, made progress in *recognizing* speech. In the early days of speech-recognition systems, they had to be taught to recognize each speaker separately, and each word had to be spoken with a distinct pause between. Now, although systems still improve as they gain familiarity with a speaker, such systems can recognize, to a limited extent, the speech of strangers. If a person speaks at a normal speed, these systems can recognize a limited one-thousand-word vocabulary. If the speaker separates his words with brief pauses, the systems can recognize up to sixty thousand different words. Such systems are used by the handicapped to operate devices with spoken commands, and by lab technicians who can speak their observations as they look into a microscope. They're even incorporated into certain cell phones: just speak the name of the person you want to call and the phone dials for you.

Threepio's speech recognition goes far beyond the English language, though. As he likes to remind us, he knows six million forms of communication, and can translate between them, such as between Artoo's robot language and English. A new software system, Verbmobil, is being developed to serve as a translator of German or Japanese into English. We might consider this Threepio's great-great-great-great-grandfather. Verbmobil's advanced speech-recognition system makes a best guess for each word you speak, checks this information against the spoken word's stress and pitch, and then analyzes the results in two different ways. Deep analysis pulls out grammatically correct strings of words and retrieves their meaning from a dictionary of such strings. Shallow analysis picks out words or phrases that have appeared earlier and compares the earlier context to the current one. Verbmobil then looks at all the possible things you might have said:

1. May the force be with you.
2. May the fours be with you.
3. May the fours bee with you. .
4. May the fourth, be with you.
5. May the force be with ewe.
6. May the force be with Hugh.
7. May the force b with u.
8. May the force be with thew.
9. May the foreseeable.
10. Mae LaForsby with Hugh?

and picks the most likely. Hopefully it's the right one.

You might think with all these capabilities that Verbmobil will soon be taking over translation at the United Nations. Unfortunately, so far it's only being programmed with the limited vocabulary needed to translate conversations about making appointments. Shall we meet at 2 P.M. for a secret attack on the Death Star?

After deciding what you most likely said, Verbmobil translates your statement into an equivalent in Japanese or German. This of course does not mean Verbmobil *understands* what was said, or its translation. It simply knows which phrases are equivalent to others.

Verbmobil then articulates the translation with a speech synthesizer. While Threepio's voice sounds quite human, current electroacoustic speech synthesizers are less successful at mimicking the intonation and rhythm of human speech. Their speech tends to be slow and monotonous, with an electronic twang.

On a basic level, it's hard enough to tell a computer how to pronounce words. Many words that are spelled similarly are pronounced quite differently, such as *have* and *gave*, *through* and *cough*. A letter is not always pronounced the same way. So a simple series of pronunciation rules will not be sufficient. In addition to general rules, the computer needs a pronunciation dictionary to pronounce each word correctly. We have computers now that have this information and can pronounce words quite accurately. But that's not enough to make the computer's speech sound human, or even natural.

Our speech varies in pitch, volume, speed, and stress. Programming in all these different qualities, so that the speech generated will sound human, would be impossibly complex. One promising approach that might allow scientists to avoid some of the difficulties of programming in all these various factors is to use neural networks to generate natural computer speech. As we discussed above, a neural network can be "trained" without any overall governing program, and it is particularly suited to recognize complex patterns, such as those in speech.

A few years ago, Dr. Terrence Sejnowski and colleagues at Johns Hopkins University created NETtalk, a neural network designed to generate speech. NETtalk was given a passage to practice reading over and over again, and was told how each word should be pronounced. The network began generating unintelligible sounds, then developed recognizable baby-type talk, and within a few hours of training learned to pronounce 92 percent of the words in the passage accurately. With current improvements in hardware, it's estimated that such training could potentially occur in only a few seconds. We might then imagine Threepio's speech generator, before it was installed, undergoing similar though much more extensive training, learning how to pronounce a wide variety of sentences and phrases in six million different languages. I wonder if Anakin did that. In the event that Threepio needs to speak a phrase outside his training, he might then use case-based reasoning to find a comparable phrase, and base his stress and intonation on that.

What is apparent in listening to Threepio is that he is not just pronouncing words; he is speaking, with intelligence, will, and understanding. This means he combines speech recognition and speech generation systems with a sophisticated artificial intelligence.

And Threepio has yet another quality to his speech: emotion. Threepio's speech is imbued with emotion, from disgust to arrogance, fear to joy. Scientists are currently studying how emotions affect the way we speak. This can vary significantly depending on the person and the situation. Programmers need to find some way to quantify this in order to create emotional synthesized speech. To generate such speech, we'd need to program a computer to discern which emotion would be appropriate for a particular utterance, and to then alter those words in a way that would convey the emotion.

Choosing an appropriate emotion suggests a computer is simply mimicking or simulating a feeling. But the emotion in Threepio's speech appears genuinely felt. When Threepio stumbles onto stormtroopers in Cloud City, he actually stutters in fear. But is that possible? Can robots feel emotion?

DO DROIDS DREAM OF ELECTRIC SHEEP?

The most amazing thing about R2-D2 and C-3PO is how human they seem. They each have strong personalities, and constantly convey emotions. Threepio is a worrier and a whiner, concerned primarily about his own well-being—his favorite refrain, "Will this never end?" To be fair, he also cares about Artoo, "Master Luke," and others. Yet that affection only comes out in the rare moments when Threepio is in a good mood. He seems constantly irritated, often venting his emotions by insulting Artoo, calling him an "overweight glob of grease" and other colorful insults. He is disgusted by Jawas, self-conscious about his appearance when his legs are not attached, and generally fastidious. He views his "life" as one trial after another, and believes one small misstep will lead to the spice mines of Kessel or some other horrible fate. He is prone to melodrama; self-absorbed and insensitive to others; yet can be quite the kiss-up when necessary, as when convincing Luke's Uncle Owen to buy him or when translating for "his high exaltedness" Jabba the Hutt. Don't get me wrong; I love Threepio. He's anything but a typical hero. And as a robot, he's fascinating. I just don't think I'd like to be locked in a trash masher with him.

Artoo, even without the ability to speak English, manages to convey a clear personality himself, and to express a range of emotions. Artoo is loyal to the humans he serves, and he cares about them, for example when he stands out in the cold of Hoth scanning for Luke. He also cares about his counterpart Threepio, though Threepio sometimes irritates him, driving Artoo to call him a "mindless philosopher" and to give Threepio the raspberry. He doesn't like being alone, as when Threepio separates from him on Tatooine, and he's frightened when he comes in contact with the Jawas. Artoo can be excited, as when he discovers Princess Leia is a prisoner on the Death Star; embarrassed, as when he falls from Luke's X-wing into the swamps of Dagobah and whistles casually to cover; and he can be stubborn, as he is when Yoda tries to take a small flashlight from Luke's camp.

Could computers and robots be given human-type emotions and personalities? And why would we want to give them such emotions? In science fiction, emotional computers and robots usually end up wreaking havoc. In *2001: A Space Odyssey*, the Hal 9000, which has the ability to perceive the emotions of others and express his own, kills all but one of the spacecraft's crew, and expresses fear as the remaining astronaut turns him off. In *Saturn 3*, a robot lusts after Farrah Fawcett and goes on a killing rampage. In the *Star Trek* episode "The Ultimate Computer," a computer given its creator's personality and emotions fears that it will be turned off. It believes that war games are actual attacks and begins shooting at friendly ships. When finally convinced by Captain Kirk that it has committed murder, the computer feels guilt over its mistake and kills itself.

In these cases, and in life, emotion is often perceived as negative. Too much emotion in a person or a robot can lead to irrationality or psychosis. Decisions made out of emotion are considered unwise. Most people believe robots should be rational, logical, and scientific, unaffected by emotion. Yet some scientists argue that while too much emotion can cause irrational behavior, so can too little.

Let's first look at how decisions are made in the human brain. We might like to think that we weigh options logically and unemotionally, but scientists are finding that this is not true. Researchers have spent a lot of time trying to pinpoint the different sections in the brain where abstract thinking and emotional responses occur. The neocortex, made up of gray matter, forms the outer layer of the two large hemispheres of the brain, and is believed to be the location of most thought. The limbic system, the more interior section of the brain that includes the hypothalamus, the hippocampus, the amygdala, and the anterior cingulate cortex, is believed to be the location of emotion, memory, and attention.

Yet what scientists are now realizing is that most functions of the brain tend to involve both the neocortex and the limbic system, both logic and emotion. The systems work in concert, intertwined, information constantly passing between them. Emotions do not *intrude* on reason; they are actually a critical part of it.

In a normal human, only the simplest decisions can be made totally logically. Dr. Rosalind Picard, associate professor at the MIT Media Lab and author of *Affective Computing*, explains that it must be "a short, well-defined decision. Given this rule, you get that decision. For example, if I'm picking up trash, and I know that a soda can lying sideways on the ground is trash, then when I see such a can, I decide to pick it up." Computers are quite good at making these kinds of decisions. More complex human decisions, however, require emotion as well as logic. Yet we expect computers to make these decisions using only logic. Perhaps computers have been unable to reason intelligently because they are missing the equivalent of the "emotional" part of a brain. After all, if emotions and desires served no purpose, why would they have evolved in us and in so many animals? Recent research reveals that emotions help motivate us; help us set priorities; help guide our reasoning, planning, and decision making; help us focus our attention; and help us cope with adversity.

Evidence of the importance of emotions has been found by Dr. Antonio Damasio, M. W. Van Allen Professor of Neurology at the University of Iowa College of Medicine and author of *Descartes' Error*. Dr. Damasio treats patients with frontal lobe disorders. In these patients, communication between the neocortex and the limbic system is impaired, giving Dr. Damasio an opportunity to study how humans function when logic and emotion are not intertwined. The patients seem extremely logical and intelligent, yet unemotional. As Dr. Damasio says, they seem "to know but not to feel."

You might think such people would act very rationally. And sometimes they do. Dr. Damasio relates the story of one patient who drove to the doctor's office over treacherous roads after an ice storm. With complete calm, the patient navigated the slick roads, utilizing the appropriate methods for dealing with icy conditions. The woman driving in front of him skidded on the ice, reacted inappropriately by braking, and spun into a ditch. While most of us would have reacted to her skid in fear and perhaps hit the brakes ourselves, the patient had no such reaction, continuing sedately on his way, following the correct procedures. Here, the patient's unemotional state worked to his advantage, much like we imagine a computer's unemotional state to work to its advantage.

Yet in actuality this lack of emotion is more often a hindrance than a help. As emotion declines, so does the power to reason. Dr. Damasio's logical patients are actually unable to make rational decisions. To make even the simplest decision, they consider each possible option, debating endlessly with themselves about which is best.

Dr. Damasio relates another story of the same patient. The doctor suggested two possible dates for their next appointment, and the patient then spent nearly a half hour discussing the pros and cons of each date, covering every possible circumstance, factor, option, and consequence: other appointments, other commitments, the cost of gas, the weather, and on and on. The doctor finally cracked and picked a date himself.

Most of us, after a brief indecision, would choose an option. We would decide one factor outweighed the others, or if there was little difference, we would go with a gut feeling that one option was better than the others, or we would choose something at random. We would know that any prolonged consideration of this minor issue would be a waste of time, and embarrassing as well. Those negative emotional associations would keep us from debating the issue at such length. Yet those negative associations did not play a part in the patient's actions. In this case, the patient's unemotional state crippled his ability to make a simple decision. "Even though emotions are considered quite primitive phenomena," Steve Grand says, "they are clearly a very important aspect of intelligence."

If a robot is flying your spaceship, you certainly don't want it "slamming on the brakes" in a panic when this could create a dangerous condition. Yet you also don't want it endlessly debating whether it would be better to visit Obi-Wan on Tatooine first and then go to Coruscant, or vice versa.

And indecision isn't the only problem created by this lack of emotional input. People with frontal lobe disorders tend to repeat the same bad decisions over and over. Dr. Damasio tested patients with a gambling game designed by his postdoctoral student, Antoine Bechara. A patient was "loaned" $2,000 of play money and told to lose as little as possible and to try to make more. Four decks of cards were laid out on the table. The player was to turn over one card at a time, and a monetary reward or penalty would result. The players were not told how the rewards or penalties were decided. Two of the decks provided rewards of $100 interspersed with high penalties of as much as $1,250. The other two decks provided rewards of $50, interspersed with much lower penalties, no higher than $100. "Normal" players would experiment with all four decks, but then quickly realize that the higher-paying $100 decks were too dangerous, carrying them near bankruptcy. They would then stick to the $50 low-risk decks, turning many more cards in these decks.

Players with frontal lobe disorders turned many more cards in the "dangerous" $100 decks. With the game only halfway over, they would often have lost all their money and need to borrow more. Yet even after borrowing more, they persisted in their previous pattern of behavior. The negative outcome did not deter them.

We would associate the dangerous decks with bad feelings and the safer decks with good feelings, and so we would be drawn back to the safer decks. In addition, most of us would feel shame over making a bad decision, which would focus our attention on avoiding a repetition of the mistake. The patients don't make emotional associations with each deck, and they don't feel shame over making a bad decision. Thus, they aren't deterred from repeating that bad decision. This causes problems not only in card games but in life. The patients repeat the same mistakes there as well, losing money in poor investments, starting up ill-conceived new businesses, marrying unsuitable mates, and more. Dr. Damasio concludes that feelings are "an integral component of the machinery of reason."

One additional problem has been documented in those with frontal lobe disorders: the inability to remain focused on a goal. While most of us set a hierarchy of priorities and attach different emotional urgencies to different tasks, Dr. Damasio's patients are unable to do so. Another patient, Elliot, found that he could not keep focused on a single task or goal. If Elliot was given a pile of documents to sort, he could easily become involved in reading one and spend hours on that, distracted from his task.

One then can argue that computers without emotions are prone to the same problems. Indeed, artificial intelligences have difficulty focusing and setting priorities. They also tend to make the same decisions over and over, whether those decisions lead to good outcomes or not. And they are not good at coming to those decisions. Just like people with frontal lobe disorders, they can become overwhelmed by the number of possible options and waste time in an exhaustive consideration of every factor. For example, in *A New Hope*, Artoo must decide whether to stay at Luke's house or to seek out Obi-Wan. Yet there are more options than this. Artoo can stay at Luke's house and tell the truth about his secret mission; he can stay at Luke's and attempt to send a message to Obi-Wan; he could stay at Luke's, forget about his secret mission, and take an oil bath; he could leave for Obi-Wan's at night; he could wait until morning to leave; he could have Threepio load him into a speeder to reach Obi-Wan's faster. And on and on. While humans might instinctively dismiss fifty out of sixty options, computers have a difficult time duplicating this type of elimination process. Dr. Picard says, "They can't feel what's most important. That's one of their biggest failings. Computers just don't get it."

Dr. Picard believes that emotions could remedy some of the most striking failures of current computers. Emotions could potentially aid in the decision-making process, associating certain options with good or bad "gut" feelings, and giving the computer a sense of the importance of various factors. Emotions could help computers realize certain actions or decisions lead to negative results, and avoid repeating those mistakes in the future. Emotions could help robots set priorities, create motivations, make decisions, focus their attention, and communicate more helpfully with humans. This last benefit, which relates to one of Threepio's main duties, explains why Threepio's creators would have wanted to provide him with some emotional ability. Both Threepio and Artoo seem able to set priorities, remain focused and motivated, make decisions, and interact rather effectively with humans, which suggests that they do have emotions.

If we did want to give a computer or robot emotions, exactly what would we need to do? Dr. Picard proposes that computers should be given the ability to recognize, express, and even feel emotions. Let's take these one at a time. First, we'd need to give the robot the ability to sense the emotions of others.

WHY HAN AND THREEPIO WILL NEVER BE FRIENDS

If a robot can recognize our emotions, it can use that information to guide its decisions and behaviors, making it more useful to us. The ability to recognize emotions would tell a robot what is important to us, what is irrelevant, what actions or data satisfy us, and what actions or data leave us dissatisfied. If my Microsoft Word program could read my emotions, it could tell when its "help" function has failed to help me and I'm bursting with frustration—which happens just about every time I use the "help" function. In such a case, it could adjust its behavior, offering me additional options or simpler instructions. If it found that I remained frustrated even with these additional options, a simple "I'm sorry" message would be nice.

How can we teach computers to detect our emotions? We reveal emotions in a number of ways: visually, through our expressions, posture, and gestures; aurally, through the inflection of our voices; and through various bodily processes, such as heart rate, temperature, and blood pressure.

Earlier we discussed how difficult it is to make a computer understand visual input. If we are successful at teaching our droid to "see," then we might have it study the human face for patterns that correspond to different expressions. Facial expressions tend to reveal whether we feel positively or negatively about something. In addition, researchers have connected certain facial muscles and movements with specific emotions, so we could potentially program such information into our droid. Systems are now being developed to distinguish a smile from a frown, or to track eyebrow movements and equate these with emotions.

Yet people express emotions differently. Luke expresses his disappointment at not being allowed to go to the academy by hanging his head and frowning, while Ben expresses his disappointment at Luke's insistence on leaving Dagobah for Bespin by simply closing his mouth. These expressions can vary depending on age, sex, the culture one is raised in, and the specific situation. To make life even more complicated, sometimes people hide or fake emotions. For example, on Cloud City Lando hides the extent of his anger at Darth Vader, so that his plans to double-cross Vader will have a greater chance of success. To make matters more difficult, emotions vary in intensity and often mix with each other to create unique, unnameable emotional states.

A robot with diplomatic duties like Threepio would need to be able to recognize different emotions not only in humans, but also in a variety of alien species, each of which would have its own ways of expressing emotion. Wookiees, for example, have rather inexpressive faces yet reveal emotion through voice and body language. If you've ever had a pet, you've gone through the process of learning to "read" that animal's emotions. I swear I can tell exactly what my iguana is thinking, but it took me several years to develop that ability. Threepio would need this skill six million times over. He even seems able to recognize emotions in other robots. He knows that Artoo is upset when Luke is missing on Hoth.

Taking all this into consideration, we can see it would be quite difficult to give robots the ability to detect emotions. One way to make this task more manageable would be to have the robot focus on learning the emotional states of just one person, its owner. This would make it much easier for the robot to learn which expressions correspond with which emotional states. Just as we're best able to "read" the emotions of those closest to us, so would droids. Dr. Picard notes that when Threepio is not engaged in conversation, he often looks at Luke, then looks at whatever has Luke's attention, then looks back at Luke. "Threepio would constantly look at Luke's face, even though Luke wasn't talking to him. What was Threepio doing then? It was almost as if he was reading Luke's face and watching for signs of approval or disapproval, seriousness or distress." In this way, Threepio could learn which expressions are usual for Luke in various situations. "You have to watch someone a long time to understand what their expressions really mean," Dr. Picard explains. A droid could combine its visual input with other information about its owner, such as habits, preferences, personal goals, reactions and expressions recorded in previous situations, in order to better judge its owner's desires, and so better perform its job. To judge the desires of strangers, a droid might then compare the strangers to its owner to try to understand them.

If our robot can also "hear," we might teach it to study the intonation of our speech to deduce our emotional state. Tone can be very revealing of emotion. Though my iguana, Igmoe, makes no sounds himself, he can tell when my tone is reassuring, scolding, and even impatient, and as he detects these various tones, his response varies accordingly. As with many of the qualities we've discussed thus far, programming this information into a robot is quite complex. If the content of a person's speech is emotionally neutral (meaning the person is saying something like "Open the door" and not "Just open the door, you stupid lump!"), people correctly identify the emotions of the speaker only 60 percent of the time. So how can we expect robots to do better?

Scientists studying how speech varies with emotion are discovering that many qualities are involved, including pitch, loudness, inflection, articulation, and rhythm. Dr. Deb Roy and colleagues at MIT have created a program that can successfully distinguish between an approving voice and a disapproving voice about 75 percent of the time. With some speakers the emotion is much more reliably detected than in others, since some people's voices are more expressive than others. My research assistant, Keith Maxwell, has a Joe Friday–rapid-fire delivery that reveals very little emotion. Yoda's voice, on the other hand, is extremely expressive.

For Threepio to be an effective translator, he needs to be able to recognize the emotion with which someone speaks, translate the speaker's statement into another language, translate the emotion into the equivalent inflection, rhythm, and pitch in another language, and speak the translation with this particular intonation.

Other sensory data might refine the robot's emotional acuity. While the voice tends to reveal the intensity of emotion and facial expression to reveal whether the emotion is positive or negative, additional sensors could help even more. Mood rings popular in the seventies claimed to reveal your mood by changing color with your body temperature. If Threepio had the ability to measure temperature, heart rate, respiration, blood pressure, and pupilary dilation, like a sophisticated lie detector, he could be a more reliable emotion detector. Combining this data with information about the individual and the situation could provide a higher degree of accuracy. Yet such abilities could also provide conflicting data. Say Threepio is attempting to deduce Leia's emotional state. Her heart is racing, her blood pressure is up, her face is tense and tightly controlled. Threepio might conclude that she is very upset, as she is when she comes face to face with Darth Vader on Cloud City. Yet she exhibits the same characteristics after kissing Han on the *Millennium Falcon*. There her emotional state is quite different.

If *Star Wars* robots did have the capability of reading emotions, though, such information could certainly be useful in diplomatic negotiations and in rebel planning. A droid would be able to tell whether someone was being truthful or planning a trap. In such a case, the big winner would be Vader. With a mask covering his face, his voice enhanced and altered, and his respiration and other bodily processes regulated by machinery, it would be very difficult to read his emotions.

Using the techniques discussed, computers may in the near future have the ability to recognize human emotions. But do Threepio and Artoo have this ability? For Artoo, this question is difficult to answer. He actually seems to rely, at least in part, on Threepio to interpret human emotions for him. The most revealing evidence comes early in *A New Hope*. Artoo plays part of Princess Leia's holographic message in front of Luke. He tricks Luke into removing his restraining bolt, then denies all knowledge of the message and fakes technical difficulties. Luke, angry and frustrated, is called away to dinner. Threepio tells Artoo he better play the message for Luke, and Artoo beeps back a question. Threepio answers, "No, I don't think he likes you at all." Artoo's question apparently is, "Does our new owner like me?" He seems to have an understanding of emotions and a sense of their importance, yet he may be unable to evaluate human emotions himself, or at least he may recognize Threepio's superior abilities in this area.

Threepio here reveals a sophisticated ability to detect emotions. Through Luke's angry tone, sharp emphatic gestures, and perhaps other signals, Threepio has deduced that Luke is not pleased with Artoo's behavior. Thus Threepio has an important function, not only translating words between humans and droids, but conveying emotions. Artoo can then discover how well he's succeeded at pleasing his master.

Threepio is also watchful for signs that he is pleasing his master. When Threepio discovers Luke is giving him as a gift to Jabba the Hutt, he puzzles over Luke's motivation, saying Luke "never expressed any unhappiness with my work." Obviously Threepio has been monitoring Luke's level of satisfaction with him.

Another example of Threepio's ability to detect emotions occurs when Luke is missing on Hoth. Threepio, translating Artoo's beeps, tells Leia the chances of survival. After a moment of hesitation, Threepio adds that "Artoo has been known to make mistakes . . . from time to time." He has realized, albeit belatedly, that Leia is upset and telling her the long odds has only increased her concern. Not only does he detect her emotions, he tries to change them and make her feel better.

Threepio's abilities, though, are far from perfect. In *A New Hope*, when he hears Luke and the others crying out in joy that the trash masher has stopped, he incorrectly interprets their cries as screams of pain. Part of the reason for his error may be that he has only the poorly transmitted sound from the comlink to judge by. Dr. Picard comments, "What's fascinating with that example is that Threepio is making the same mistake that our state-of-the-art speech affect analysis makes." While computers can deduce the level of excitement or intensity in someone's voice, that can't yet reliably tell whether that emotion is positive or negative. Threepio here mistakes a positive emotion for a negative one. Without visual or other input, the mistake is one a human might well make.

Threepio's biggest failure to read emotions occurs with Han Solo. If Threepio understood how his complaints and protests irritated Han, he could attempt other methods of interacting with Han. But Threepio has Han close to bursting a blood vessel through most of *The Empire Strikes Back*. He irritates Han to the point that Han has Leia turn Threepio off! Certainly a robot who can detect emotions should know enough to avoid this dire situation.

Yet let's look at things from Threepio's point of view. He's apparently used to dealing with Princess Leia and other diplomats.

And he seems to function fairly well with Leia and with Luke, who treats him with respect. He may never have met anyone like Han Solo before, and so have trouble deducing Han's emotional state. Han expresses himself much differently than Leia. He's the king of sarcasm, and often says one thing when he means exactly the opposite. Dr. Picard recognizes this could be a problem. "Humor and sarcasm involve more than just recognizing a face or a voice. They also require some situational understanding and common sense."

A failure to recognize Han's tone could easily lead a droid to misunderstand the meaning of what was said. Sometimes Threepio recognizes Han's sarcasm, as when he points out to Han that the asteroid they've landed in is unstable. Han replies, "I'm glad you're here to tell us these things." While the words themselves praise, Threepio clearly realizes that he's being criticized and takes offense. Yet later, when Threepio interrupts Han and Leia's kiss to announce the exciting progress he's made in repairing the hyper-drive, Threepio mistakes Han's sarcastic "Thank you. Thank you very much," for sincere praise. His ability to read sarcasm—or at least Han's sarcasm—isn't perfect.

In Threepio's eyes, then, Han is erratic: sometimes rude, arro-gant, and insulting; sometimes helpful and appreciative. If Three-pio is attempting to please, these mixed signals could generate conflicting impulses within him. No wonder Threepio calls Han "impossible."

Yet misunderstandings aren't the only roadblock to a friend-ship between Han and Threepio. Han clearly has a general dislike of droids. If Threepio is programmed to serve his owner—and his owner's friends—in a satisfactory and pleasing way, he's going to have a long wait for his actions to bring a smile to Han's face.

Even if Han didn't dislike droids, Threepio might still have a hard time making Han happy. Not only does he need to detect Han's emotions; he needs to know how to respond suitably. "Knowing how best to respond," Dr. Picard says, "may be a much harder skill for us to give robots."

One reason humans have some ability to respond suitably to the emotions of others is that we have emotions of our own. This allows us to understand the feelings of others, or sympathize. Dr. Picard at first believed robots shouldn't have their own emotions.

"I wasn't sure they had to have emotions until I was writing up a paper on how they would respond intelligently to our emotions without having their own. In the course of writing that paper, I realized it would be a heck of a lot easier if we just gave them emotions." For that to be the case, the robot must actually have emotional reactions to what's going on around it.

As we discussed earlier, both Artoo and Threepio have clear emotional responses to events and people. Artoo feels strong loyalty and friendship for both Luke and Threepio. Threepio constantly describes himself in emotional terms, saying he's "embarrassed," "sorry," and "afraid," and he even anticipates the emotional consequences of an event: "I'm going to regret this." Even the tiny box-on-wheels droid on the Death Star appears freaked out when it runs into Chewbacca, letting out a shriek and turning tail. But can we create a robot that feels pride, fear, frustration, and affection?

I'M OKAY, YOU'RE AN "OVERWEIGHT GLOB OF GREASE"

In humans, emotions arise in several ways. They can arise from chemicals in our bodies, from the way we carry our bodies—smiling can make you feel happier—from sensations, such as hunger or pain, or from thoughts.

Just as a person seldom makes an entirely logical decision, a person seldom has a completely emotional reaction. There are a few exceptions, instinctive emotional reactions that in essence hijack our bodies and occur without any thought at all. For example, if you see a large object flying at you, you'll immediately feel a fear response and jump out of the way before you have a chance to think. Such a response could be very useful in a droid, so that it wouldn't get hit by stray speeders or blaster fire. In fact, Threepio displays such a response when he and Luke are attacked on Tatooine by Sand People. As a Sand Person leaps at Luke, Threepio's eyes light up and he jerks backward, falling over. This seems to be an instinctive rather than logical response. If Threepio had stopped to think, he would have realized that he was standing at the edge of a cliff, and that falling could cause him more damage than the Sand Person. In fact, the fall rips his arm out of its socket (much as an angry Wookiee might do).

Most emotions arise more slowly, in combination with thoughts. For example, a person on an unfamiliar desert planet could grow more and more concerned as he walks farther and farther and finds no food, water, or shelter. As he realizes his resources are dwindling and he may die before he reaches help, he will begin to feel fear. A droid in the same situation, like Artoo on Tatooine in *A New Hope*, may similarly recognize that his energy levels are falling and that he may not reach help before his energy supply is exhausted. In such a situation, the droid may also feel "fear."

Why would you want a droid to feel fear? Dr. Picard explains that in this state, the droid's priorities and behaviors could change, just as a fearful person's priorities and behaviors change. A person will use adrenaline to push himself to his limits. He'll become more watchful for threats and useful resources, and he'll focus all his attention and energy on survival. A droid in the fear state could access its emergency power supply; channel additional power to its sensors to watch for signs of life or dangers; shut down nonessential systems; and focus on survival over any secret mission or other task. Artoo displays additional watchfulness just before he's captured by the Jawas. We might even imagine that he rolls into the Jawas' trap on purpose, his need for an energy recharge overruling his instructions to complete his mission quickly and secretly. In such dire circumstances, the positive value of an energy recharge far outweighs the negative value of capture. Emotions in robots can thus alter priorities, change behaviors, make judgments of value, and help make decisions flexibly, quickly, and well.

Similarly, Threepio exhibits a change in priorities during the holographic chess game between Artoo and Chewbacca aboard the *Millennium Falcon*. At first Threepio seems to want Artoo to win. But when Threepio learns that Chewbacca may become violent if he loses the game, Threepio's priorities change. His top priority is now to avoid angering the Wookiee further: "Let the Wookiee win." If Artoo continued to play at the same level of ability, he and Threepio might have ended up in a very unpleasant situation.

Fortunately, the danger triggered a change in them equivalent to a change from a normal state to a fear state.

Yet if we enter the holographic chess scene earlier, we see that Threepio's change does not occur immediately. During the chess game, Chewbacca howls when Artoo takes one of his pieces. Threepio complains, "He made a fair move. Screaming about it can't help you." Threepio recognizes that Chewbacca is expressing frustration and dissatisfaction. He may be deducing this from Chewbacca's body language, his words, or his tone of voice. Yet at the same time he is dismissive of Chewie's feelings. His dialogue suggests Chewie should behave logically, not emotionally.

Threepio resists switching into a fear state, preferring to maintain his original priorities and goals rather than to forego them in reaction to Chewbacca's feelings. Just like a person, Threepio values his own feelings over anyone else's. A selfish robot? Yes, and this quality is necessary if Threepio is to carry out preassigned tasks. If Threepio valued everyone else's feelings above his own, he'd spend all his time trying to make everyone happy, the ultimate codependent enabler, rather than focusing on his own goals and duties.

As it is, Threepio tries to make his owner and his owner's friends happy, while at the same time operating in the way he sees proper and accomplishing his tasks. Although he is able to recognize others' emotions, his recognition of them often doesn't change his own actions, rather like an insensitive human. The other party's emotions must pass a certain threshold for Threepio to change his goals. Dr. Picard agrees. "You don't want a computer changing its behavior every time you twitch." She also points out that Threepio only changes his behavior when their mission is jeopardized. "Artoo is critical to their mission. If the Wookiee damages Artoo, that's going to threaten getting the information in Artoo to the people they need to get it to."

When Han explains that it is unwise to upset a Wookiee, Threepio replies, "But sir, no one worries about upsetting a droid." This line provides a key to understanding Threepio. He obviously feels organic beings are insensitive to the feelings of droids, and that they have upset him many times. Yet, in a very human way, he can be staggeringly insensitive to and dismissive of the feelings of organic beings. It's only when his safety is threatened—when he fears Chewie will rip them apart—that he decides to take Chewie's emotions into account. Threepio's top priority is, after all, saving his own neck.

So a droid must not switch emotional states too easily, yet it must switch when the situation warrants it. Since *Star Wars* droids are so emotional, for an example of a failure to switch states, we actually have to turn to a human character. If humans repress and ignore their emotions, they can make mistakes similar to those a computer without emotions might make. When Grand Moff Tarkin is told the battle plan of the rebels poses a real threat to the Death Star, he does not switch to a "fear state," which would be appropriate in this situation. Instead he maintains his previous priorities—destroying the rebels, proving the superiority of the Empire—which are inappropriate in the face of this new information. The failure leads to his death. (Threepio would have surrendered in a flash.)

So emotions can be very useful for a robot. The particular emotions that might be most helpful for a particular robot depend on that robot's function. Dr. Picard believes that within the next year, we'll have computer scientists calling themselves "personality engineers" who will consider such issues. Actually, Steve Grand, in creating the computer game "Creatures," functioned as a personality engineer, giving his creations desires and emotions, like hunger, loneliness, and anger. His goal was "to create small furry creatures (Ewoks were in my mind when I started) that people would enjoy keeping as pets." These desires and emotions help motivate the creatures to learn and engage in lifelike behavior, which keeps their human owners attached and involved in their lives.

For a protocol droid, a love of all things proper and a dislike of inappropriate conduct might be helpful. Threepio certainly displays this. He seems to hold Chewie in contempt for whining about the chess game, and he despises Jawas for their filthiness. If a droid is meant to perform complicated missions by itself, some useful emotions would be loyalty, fear, and determination: emotions we see in Artoo and Threepio. If a droid is meant to interact with people, an emotion like "affection" would be useful. If you feel affection for someone, you want to do things that make that person feel happy. A robot, then, would want to do things that make its owner happy, that make its owner "like" it. Earlier, we noted that both Artoo and Threepio are concerned with pleasing Luke. Any behavior that displeased Luke would carry negative associations that would deter the droids from repeating it, while any behavior that pleased Luke would carry positive associations that would encourage the droids to repeat it.

When Threepio has Artoo shut down all the Death Star's garbage mashers before his owner, Luke, is squashed, Threepio feels good. The behaviors that led to this outcome—remembering to turn on the comlink, working with Artoo to solve the problem, taking a global action rather than a specific one (shutting down all the garbage mashers rather than just those on the detention level), saving his owner from danger—would all be reinforced, connected with feeling good. So Threepio would be more likely to perform such actions in the future.

In addition to caring whether Luke likes him, Artoo is even concerned with whether Threepio likes him. After Artoo fails to play Princess Leia's message for Luke, and Threepio tells Artoo that Luke doesn't like him, Artoo beeps another question. Threepio answers, "No, I don't like you either." Since Artoo and Threepio are "counterparts," it would make sense to program them with "affection" for each other. If one droid is giving output that the other doesn't find useful, or doesn't "like," their partnership will not be very effective. In fact, this happens on occasion, as in the situation just described, which leads to Artoo and Threepio's affectionate bickering. They seem to be in a constant negotiation, each trying to make their relationship more pleasing or helpful for himself. Artoo tries to convince Threepio to follow Artoo's priorities, and vice versa. Dr. Picard points out that Artoo seems to carry more authority than Threepio. Threepio will follow Artoo when he has no idea what Artoo is doing, as when he joins Artoo in an escape pod at the beginning of A New Hope and drops to the surface of Tatooine. Yet Artoo will not follow Threepio unless he agrees Threepio's course of action is the correct one. Perhaps this is because Threepio is "younger" than Artoo.

Since Threepio and Artoo are programmed to work together, to try to help and please each other, then Threepio's sticking his neck out for Artoo is quite natural, and not at all beyond his capacity. While they are occasionally unable to cooperate—as when they split up on Tatooine—they are usually able to work together to satisfy both their goals. This creates a bond of trust, loyalty, and friendship between them. Steve Grand is now embarking on a project with a colleague to build "something as close as we can get to a robot like Artoo." Yet he has decided that rather than trying to build one robot, he should build twins. "Human twins tend to bond strongly, understand each other well and develop shared language and a shared understanding. The way Artoo and Threepio interact is just like a pair of twins. I'll bear them in mind for inspiration!"

Not all droids feel affection for each other. When Artoo reports that the Cloud City computer told him the hyperdrive on the *Millennium Falcon* has been deactivated, Threepio berates Artoo for trusting a strange computer. Just as Artoo and Threepio would be programmed to try to work with each other, they would be programmed with distrust and secretiveness toward others, for security purposes.

If two droids are programmed with compatible goals, one should be able to make the other like it. But what happens when a droid is unable to make its master like it? In *The Empire Strikes Back*, Threepio warns Han of the horrible odds against successfully navigating an asteroid field, attempting to save him from danger. He is most likely repeating a behavior that has been rewarded in the past, and one that is deeply ingrained in his programming. Yet Han does not respond with happiness, as Threepio expects. This would make Threepio "feel bad." He would connect this behavior to negative results, and would be motivated to come up with alternate behaviors to use in future situations.

Yet with Han, nothing Threepio does allows him to "feel good"—except when Threepio misunderstands Han's sarcasm for sincerity. And his programming most likely forces him to inform humans when they engage in dangerous activities. Threepio's frustration and distress would grow, since he can't avoid feeling bad. When a person feels bad, the way he perceives things changes. He feels it more likely that bad things will happen. This helps to explain Threepio's negative outlook, his belief that life is one horrible trial after the next, which seems particularly strong in this movie. To add to his stress, Threepio's life and the lives of his "owners" have been in danger since the Empire's forces arrived at

Hoth. He is scolded for pointing out the dangers and is unable to do anything to stop the danger himself. His behavior grows ever more frantic and desperate until he finally suggests surrender, and Leia turns him off.

Dr. Picard offers two suggestions. First, Threepio "should find a good time to interrupt and say, 'Han, I need some feedback from you. Every time I present you with these statistical odds, you get really irritated at me. Should I simply stop doing that?' " Of course, this solution requires Han let Threepio speak three consecutive sentences before shutting him up. And that Han will not respond sarcastically, "No, Threepio, I love it when you tell me the odds." Dr. Picard's second suggestion—have Han sell Threepio or "take him back and have his personality reengineered. Reengineering a human personality, though, takes a lot of work, years of therapy, and you don't know what you'll end up with. It may be that way with robots too."

Just as feeling bad affects a person's outlook, so does feeling good. A person who feels good tends to see the world "through rose-colored glasses," interpreting everything in a positive light. We see a brief glimpse of this when Threepio excitedly reports to the kissing Han and Leia the progress he's made in repairing the hyperdrive of the *Millennium Falcon*. Threepio, in his rush of positive feelings, mistakes Han's sarcastic thanks for sincere praise.

We talked at the beginning of this section about the chemical and thought processes that create emotions. We've shown how thoughts can potentially lead to emotions in computers. But do computers need the equivalent of a chemical system to truly feel emotions? Chemicals do play an important role in human emotions. In fact, people with chemical imbalances will feel emotions, such as chronic depression, unrelated to events in their lives. If we are to re-create human emotions in robots, then, don't we need a chemical equivalent? This remains a point of dispute among scientists. Dr. Picard believes robots can be emotional without any biochemical equivalent, and adding such an equivalent will not give robots the same feelings humans have. A robot's emotions will always be different than a human's, "because it does not have the same biochemical and sensory apparatus that we have. As long as we have different bodies, we will have different 'feelings.' "

Other scientists believe emotions depend on neural-chemical

reactions and require chemical agents. Dr. Rodney Brooks, director of the Artificial Intelligence Laboratory at MIT, has built a robot called Cog, which has certain desires programmed into it. To create these desires, Dr. Brooks has programmed in internal rewards. "We have computational simulations of endorphins and hormonal levels," he explains. Certain activities bring Cog a virtual chemical reward, and so it desires doing those activities since they will bring further rewards.

Steve Grand wanted to populate the computer game "Creatures" with creations that would seem to players to be alive. In fact, Grand nervously admits, "It's not really true that I designed them to give the illusion of life; I actually tried to *make* them alive." To achieve this end, Grand combined a neural network with a virtual equivalent of biochemistry. "I created computer simulations of neurons, biochemicals, chemoreceptors." Signals designed to mimic biochemical signals in the body are released under different conditions, strengthening certain desires and patterns of behavior while weakening others. Earlier we talked about how neural networks can be trained by adjusting the strength of their various connections. Here, the virtual chemicals do the training, strengthening certain pathways and weakening others. So if the creature does something foolish and injures itself, the "pain" it feels will trigger it to weaken the pathway of connections that led to that action.

Whether through thoughts alone or through a combination of thoughts and virtual chemicals, future computers and robots may feel emotions much like you and I do. Emotions can potentially make computers much more sophisticated and helpful. The decision-making abilities, flexibility, and independence we observe in Artoo and Threepio are all contingent on their ability to feel.

In addition to helping computers function more capably, emotions may have other consequences as well. You may have to stop cursing at your computer. You don't want to hurt its feelings.

I WHINE, THEREFORE I AM

A robot may be a seething cauldron of emotions, but if it has no method of conveying those emotions, we may never know it. If emotions make a robot a more efficient decision maker, a more flexible thinker, and a more helpful assistant, why should we care if the robot can convey its emotions to us or not?

This ability can be useful in several ways. First, we humans are used to communicating with emotional beings, and we actually prefer it. In a study by Dr. Tomoko Koda at MIT, people played two different versions of electronic poker, one with an animated face that displayed different expressions as the game progressed, and another one with no face. When surveyed after their games, a majority of people preferred playing with the animated face. They were simply more engaged in the game and found the face likable. In another study, it was even found that people preferred interacting with an animated character of their own ethnicity.

If we are to interact regularly with a robot, it may be helpful to have the robot appear humanoid, and to appear to convey emotions. In my Microsoft Word program, the help icon is a small, animated paper clip, Clippit. Clippit blinks, sways, and looks back and forth as I write this book. When I open a file, she raises her eyebrows in surprise. When I command the program to search for a word in my text, she contorts into a fiercely concentrated expression. While Clippit doesn't actually feel emotions in the sense that we discussed, she does simulate emotions, and that simulation makes me feel a bit more charitable toward her when her "help" fails to be helpful.

Even more than making us patient with software, the ability to show emotions can be used to manipulate our emotions. My research assistant, Keith Maxwell, came in one day with his fiancée's digital pet, or Tamagotchi, as they were originally called. He was under strict orders to feed, care for, and play with the tiny computerized turtle image. When she first gave it to him, he says, "I thought it was kind of stupid, because I thought who cares about a little beeping toy. You don't get any return from it." As he cared for it, though, he was surprised to find that playing with the "silly thing" was actually pleasing and rewarding: "Every day it would have a birthday, and it would get a little bigger, and its weight would go up a little. So it was interesting to watch it grow." The image has been designed to make us feel affection for it, to actually manipulate our emotions. The technology is limited, though, and so is its success. As Keith says, "I guess I cared about it for a day or two. Then I grew to hate it. It was always so irritating, always beeping." Rather like Han's reaction to Threepio, I'd say. Luckily his fiancée took back the pet before any violence ensued.

Since digital pets might not be lovable enough, more cuddly electronic pets have been developed. Plush and huggable, Furbys are five-inch-tall cuddly balls of fur that laugh, dance, and make different facial expressions. Their vocabulary of more than two hundred words allows them to interact with owners, and even to tell them, "I love you." With their big eyes and floppy ears, they've been designed specifically to be lovable. Yet apparently some of these carefully designed high-tech toys rub owners the wrong way. On the Internet you can find Furby guts graphically exposed on the Furby Autopsy Home Page. So a computer that displays emotions may gain our loyalty and affection, yet it may also arouse our anger.

ARE YOU AND YOUR COMPUTER MAKING A LOVE CONNECTION?

Steve Grand reports that the nearly one million players of his game "Creatures" experience a wide range of emotional reactions, many of them responding to the creatures as if they are truly alive. "People often grieve when their creatures die." Once, when he was showing his mother an early version of the game, "We watched as a couple of creatures engaged in a chase scene that looked remarkably like 'unrequited love,' and then the program crashed. My poor mother shed a tear when I explained that her creatures were now gone forever!"

Even Grand himself has fallen under the creatures' spell. "I was once sent a sick creature by E-mail from Australia, with a plea to cure it of a disease that was causing it to just stand there and waste away. I did some experiments and discovered that it had been born deaf and blind, and so had no idea there was a world of food out there. I figured out which simulated gene had mutated and caused the defect, and after much genetic manipulation and tender loving care, I managed to put it right. Then I mailed her back to her owners. It was only later that it occurred to me that I was just as soppy as they were, having spent a whole day worrying about the health of a data file!"

In a similar way to the Tamagotchi or Clippit, Artoo, without words, is able to convey emotions as well. His robot language of whistles and beeps appears to have been created so that humans could read some basic emotional content in it. Our feeling that we "understand" Artoo contributes greatly to our affection for him.

While Threepio's face is not expressive, his voice and his body are. His tone is quite revealing of emotion, and his gestures and body language also convey cues about his emotional state. Since Threepio is fluent in six million forms of communication, we might assume he not only knows six million languages, but six million patterns of emotional expression—intonations, expressions, and body language. He uses his ability to convey emotion to add emphasis to points he feels are important, and even to try to influence the emotions of others.

In *The Return of the Jedi*, Threepio tells the story of the *Star Wars* adventures to the Ewoks in order to gain their help for an attack on the Imperial base. Any good storyteller must manipulate the emotions of his audience, and that is exactly what Threepio does. He skillfully selects details that will call up the desired emotions in the Ewoks: fear and hatred for the Empire, affection and loyalty for the rebels. And he presents these details with intonations and gestures that reinforce these emotions.

Critical to the success of robots that display emotions is the ability we discussed earlier to detect emotions. If you are furious that your robotic maid has just thrown out your *Star Wars* videotapes, a smile on the maid's face is not going to make you feel kindly toward it. You would want the maid to acknowledge your anger and display penitence. Perhaps that's why Furbys are ending up on the autopsy table. Dr. Fumio Hara is developing a system to address this issue at the Science University of Tokyo. He has built a robot that can display six different emotions: joy, sorrow, hatred, horror, anger, and surprise. He is now at work on a program that will be able to recognize various facial expressions on humans. He plans to connect that program to the robot so that it can register the expression of a human and then display an appropriate expression in response.

If we can someday make robots more perceptive and understanding of our emotions, and more responsive to them through their own emotions and expressions, we may actually finally build the ultimate artificial intelligence: one that does not drive us insane.

THE SIX-MILLION-DOLLAR SITH

In addition to droids, *Star Wars* also features a few cyborgs, organisms with mechanical or electronic components. After Luke's hand is cut off by Darth Vader, he receives a bionic hand in its place. The hand looks like a real hand, and seems to have all the mobility and strength of a real one. In *Return of the Jedi*, Luke returns the favor, cutting off Vader's hand, and we see wires coming from Vader's wrist. Vader's hand is artificial as well. When the Emperor strikes Vader with bolts of Force energy, we see that Vader's entire arm is artificial. Perhaps his legs are as well; we can't tell. Are such artificial limbs possible?

While our technology is a long way from the bionic limbs shown in *Star Wars*, prosthetic limbs that look and feel pretty close to the real thing are now available. Artificial hands can even perform simple actions. A myoeletric sensor is implanted into the user's residual limb, or stump. By flexing the muscles in the stump, the user can trigger the sensor to send signals along a wire to a motor in the limb. With training, the user can flex particular muscles to send specific commands to the limb, making a hand open, close, or turn.

Prosthetic limbs can even be outfitted with temperature or pressure sensors, so that the temperature of a hot cup of coffee in a user's hand or the pressure of the hand against the cup can be detected. The information is transmitted from a sensor in the fingertip along wires to the residual limb. In the case of temperature, the signal heats or cools a metal plate on the stump to reflect the temperature of the object being held. This allows the user to regain some long-lost sensation and can help him avoid damaging his new hand. In the case of pressure, a tingling sensation reflects the intensity of the hand's pressure against an object. Before, users had to judge by sight whether they were clutching something tightly enough, which made for many accidents. One new design actually senses when an object is starting to slip and automatically tightens its grip. People with these prosthetic hands can use them to eat, brush their teeth, and answer the phone. Sensors would allow Luke to grip his light saber firmly, and to feel the warmth of Leia's hand.

Luke's fingers, though, are capable of independent movement, unlike the artificial hands currently available. But an even more sophisticated hand is now under development by scientists at Rutgers University. In this model, finer movements of muscles and tendons in the arms are used to send signals through a computer to motors that move individual fingers. One patient testing the hand actually played the piano with it.

In addition to the hand, other prosthetic limbs are becoming much more comfortable and sophisticated. In older devices, residual limbs used to be held within liners and sockets that were either too tight, cutting off circulation, or too loose, generating friction. Users would develop blisters and infections, and suffer serious pain. Now, more comfortable liners and sockets allow artificial limbs to be worn longer, without hurting the user.

Rather than being made from heavy plastics, wood, or rubber, prosthetic legs are now made from carbon composite, the super-strong, light material from which fighter jets are made. One type, the Flex-Foot, uses shock-absorbing springs at the toe and heel to bounce from one step to the next and provide stability on uneven surfaces. Knee joints that used to operate with a mechanical hinge are now hydraulic. The flow of fluid inside the joint adjusts the swing of the lower leg to the user's stride, so it swings faster when the user is walking quickly or running. A computer chip is even being added to the knee, which will measure the user's pace and adjust the swing with a tiny motor up to fifty times per second. With such legs, users have run races, played basketball, and even climbed mountains.

Sensors also play an important role in the latest artificial legs. Three sensors near the heel and three near the toes relay information about pressure on different parts of the foot to electrodes in the user's residual limb. To the user, these tingles quickly begin to feel like sensations coming directly from the foot. Able to feel the pressure of the ground against their feet, they can maintain their balance much better.

Scientists have even figured out how to make artificial limbs that can change skin tone to keep up with a user's tan. Most limbs have a latex covering that is matched to the user's body. But if the user tans or becomes pale, the tone will no longer match. Dr. Henry LaFuente at the University of Oklahoma Health Sciences Center has designed a limb covering made of two transparent layers, one silicone and one nylon. The layers have a tiny space between them. Dye of any color can be injected into that space to change the "skin tone" of the limb. This could come in handy if Vader decides to equip his meditation chamber with tanning lights.

Beyond his limbs, we don't really know how much more of Vader may be artificial. Obi-Wan says of his former pupil, "He's more machine now than man." We do know that he breathes with the assistance of a ventilator, and his voice is artificially augmented somehow. The *Star Wars Encyclopedia* tells us that Anakin fell into a molten pit during a duel with Obi-Wan. Let's examine what sort of damage would be caused by these burns, and whether they can account for Vader's condition.

We don't know the nature of the molten pit, but we might imagine it contains fresh, hot lava. During volcanic eruptions, many people have been burned to death by lava flows. Lava ranges in temperature from 1,400 to 2,200 degrees, which means that cloth, wood, or paper would immediately ignite. And skin cells would almost instantaneously shrivel and die. Before we can figure out the exact damage to Anakin, though, we need to know how much of his body comes into contact with the molten material, which determines the extent of the burn, and how long it is in contact, which determines the depth of the burn.

One major factor is the density of the molten material. Lava is quite dense. If the pit contains molten rock or molten metal, the liquid will be much denser than a human body, which is comprised mainly of water. If Anakin gently lays back on the molten material, then, he will float, only sinking an inch or two into the lava. This will limit burns to about 15 to 20 percent of his body. If he happens to step into this pit by accident, he will sink in about up to his knees. If he drops into the pit from a great height, though— which seems a bit more likely—his momentum will carry him deeper into the lava before he bobs back up, like a bar of soap dropped into a sink full of water. How far he goes into the lava depends on his velocity when he hits the surface. Since lava is so dense, though, we might guess that he doesn't go much deeper than the surface. And the heavier density of the molten material will drive his body up quickly, limiting the exposure of most of his skin to a few moments. The density of the lava may well be what saves Anakin's life.

Since Anakin's face was not seriously burned, we can assume he didn't fall on his face. Perhaps he landed on his feet. In that case, the most serious damage would have been done to his legs, since they would have remained in contact with the lava until he walked out of the pit, levitated himself out, or was rescued. He might have struggled to balance himself with his hands, immersing his arms in the molten material as well.

The case of a foundry worker who stepped into a pot of molten metal can perhaps shed some light on what injuries this might cause. In accordance with safety regulations, the foundry worker was wearing fireproof trousers and leather boots with metal tips, which served to partially protect him from the molten metal. Anakin most likely didn't have any protective gear on. Yet the molten metal the foundry worker stepped in was 2,400 degrees, hotter than lava. The worker's boot was almost completely destroyed. Just a small shriveled-up bit was left, about the size of the sole. The worker suffered third-degree burns on his right foot and leg up to the knee, and second-degree burns on his hands and face, from spattered metal. You might expect that his leg had to be amputated. Yet it didn't.

The initial scars, called eschars, had to be treated immediately. Dr. Michael Blumlein, physician and teacher, explains that since scar tissue is rigid, if a scar goes all the way around the circumference of a limb, the eschar "can constrict blood flow to the area, leading to the possibility of gangrene." The limb can become paralyzed, numb, and cyanotic. If the doctor cuts into this scar tissue, in a process called escharotomy, the constriction can be relieved and gangrene prevented. A few days later, the worker's dead skin was surgically excised with a specially designed knife that can shave a thin layer of dead tissue off the body. Since the skin in some areas was completely destroyed, skin grafts were applied over several operations. The worker was left with minor scars on his hands and face, and more serious scarring on his foot and leg, but he was able to walk fairly well with just a cane. And this case isn't the exception. In twenty other cases of splash burns from molten aluminum at temperatures ranging from 600 to 1,300 degrees, no amputations were required.

We can also consider the cases of two geologists who fell into lava at the Hawaiian Volcano Observatory. The lava wasn't very deep, and they got out quickly. After hospitalization for their burns, they both recovered without any serious permanent injury. Exposure to molten materials, then, need not be lethal or even crippling, as long as the amount of body exposed and time for which it is exposed aren't too great.

While Anakin may land on his feet in the pit, it seems most likely that he would land on his back. At such high temperatures, the clothes on Anakin's back would almost instantaneously burn off, and the skin would follow quickly after, the proteins in it literally cooked so that what remains is rigid and dead. In a few seconds the external cartilage of the ears and the tissue of his fingers and toes would begin to burn off. With each moment, heat would penetrate deeper and deeper into the back of his head, neck, his back, and the backs of his arms and legs, burning through fat, connective tissue, muscle, nerve, and even bone. The rest of his body would be burned by the heat radiating from the lava, though less severely. As Anakin gasped in shock, superheated air, steam, and volcanic gases would burn his mouth and upper airway. If the gases were hot enough, even Vader's lung tissue might suffer thermal damage. Oddly, after the initial shock, Anakin would feel no pain. Third- and fourth-degree burns are actually painless—at least until the surgeries and treatments begin. The worse the burn, the less the pain.

If significant amounts of Vader's muscle tissue had to be removed, like the dead tissue in the foundry worker, Anakin would be left extremely weakened. Heavy scarring on the limbs could cause additional problems. Dr. Blumlein explains that scarring "can be devastating, not just to physical appearance, but to functional ability. Scar tissue can be so thick and tenacious that joints can become frozen. I've seen people who couldn't move their necks, or hands, or fingers, because of constriction from scars." Much of this can be avoided or lessened through techniques like skin grafts and reconstructive surgery. Yet Vader might actually prefer amputation, if flexible, strong bionic limbs are available as replacements.

Damage to the airway and lungs is one of the most frequent causes of death in burn patients. Yet usually if a patient survives the acute phase of the injury, his breathing problems will clear up and leave no long-term damage. If Vader suffered some damage and scarring of his lung tissue, though, he might never regain normal functioning. This might require a system that can enrich the oxygen content of regular air and pump that enriched air into Vader's lungs.

Another possibility is that Vader's phrenic nerves may have been damaged from his burns. The phrenic nerves stimulate the movement of the diaphragm, a muscular membrane that causes the lungs to fill and empty. Since the phrenic nerves come out of the protective spinal cord at the back of the neck, and Vader may have suffered serious burn damage there, we might theorize that severe heat damage destroyed the nerves and caused a partial paralysis. If this is the case, then Vader might need assistance to inflate and deflate his lungs.

Quadriplegics who suffer from a similar problem use a ventilator attached through a hole in the neck to the trachea. Ventilators force air down the trachea into the lungs. This means air enters and leaves the body below the vocal cords, never passing over them. In a more sophisticated design, a speaking valve can be incorporated into the tube, which allows air through the tube into the lungs, but prevents air from leaving by the same path. Instead, the exhaled air is forced up over the vocal cords and out the mouth. This allows the user to speak, and he can speak at any time during the breathing cycle. Vader may have a similar ventilation device. This would explain why his breathing seems independent of his speech. A person's voice with such a system can often be weak, as you may have noticed with the actor Christopher Reeve, who has a similar system. Thus to project a commanding presence, Vader would need his voice augmented somehow.

The easiest method for Vader, though, would be to simply use the Force to pump air in and out of his lungs and move his burn-damaged limbs. But I guess even a lord of the Sith gets tired sometime.

We started out this chapter asking whether robots of the future are likely to be similar to *Star Wars* droids. The truth is that recent theories and research make emotional robots like Artoo and Three-pio much more likely than they seemed back when *A New Hope* first came out. While we still have a long way to go in designing bipedal robots and robots that can truly "see," "hear," understand, and feel, robots of the future may very well resemble those built "a long time ago, in a galaxy far, far away."

What other wonders might the far more advanced *Star Wars* technology produce? In the next chapter, we're going to explore one of their greatest accomplishments: the ability to travel quickly across vast interstellar distances.

4

SPACESHIPS AND WEAPONS

You never heard of the *Millennium Falcon?*
It's the ship that made the Kessel Run in
less than twelve parsecs.

—Han Solo, *A New Hope*

W e've whooshed down the Death Star trench in an X-wing. We've navigated an asteroid field against impossible odds. We've maneuvered through dizzying battles. We've felt the vertiginous rush of the jump to hyperspace. We've outshot TIE fighters, outrun star destroyers, and destroyed Death Stars.

One of the most striking and exhilarating elements in the *Star Wars* saga is the spaceships. That opening shot of *A New Hope,* in which a star destroyer glides endlessly over our heads in pursuit of Princess Leia's ship, reveals the awe-inspiring dimensions of that vessel. And the star destroyers are dwarfed by the mother of all spaceships, the Death Star.

These ships are equipped with high-powered lasers that can disable or destroy another ship, or in the case of the Death Star, even destroy a planet. They can travel quickly over vast interstellar distances, keeping their passengers in comfort with artificial gravity. And all without stopping for a fill-up.

Star Wars vessels aren't the most glamorous ships we've ever seen; covered with bumps and nubs, towers and guns, they have a cluttered look that prompts Luke to comment about Han's precious *Falcon*, "What a piece of junk!" They also at times break down. But that only makes them feel more realistic.

But how likely are these vessels? While life in "a galaxy far, far away" is fast paced and exciting, can we imagine ways in which we might someday be able to travel quickly and easily among the stars?

186,000 MILES PER SECOND

We all know the law. Spaceships cannot travel at the speed of light. Einstein's special theory of relativity established the speed limit and revealed that just approaching the speed of light causes horrible problems.

Let's do a little thought experiment—and be prepared, you'll find several in this chapter. Theoretical examples tend to be the best way to understand the issues involved in space travel.

Han Solo blasts the *Falcon* out of Mos Eisley spaceport, and stormtroopers on Tatooine observe him accelerating toward light speed. But as he approaches the speed of light, several strange things happen. The stormtroopers focus their scanners on the ship.

The *Falcon* is now traveling at three-quarters the speed of light. But the ship's mass appears to be one and a half times what it was when the ship was landed in the spaceport. That doesn't seem right, they think. Even if he jammed the entire ship full of spices, it wouldn't weigh that much. And the length of the ship measures only two-thirds what the registration shows.

The stormtroopers compare notes for a moment, then check the scanners again. The *Falcon* is now traveling at .9 the speed of light. But now the mass is over twice what it should be, and the length of the ship is less than half.

The stormtroopers look over their shoulders, checking that Darth Vader is not nearby to witness their confusion, then check the scanners once again. Now the *Falcon* is traveling at .999 the speed of light. It's almost reached the magic number. But now the ship's mass is over twenty times as great as it was, and its length is just one-twentieth what it was. It's shrinking into nothing!

The stormtroopers, afraid crushed necks may be in their futures, pull out the super-duper scanner, for use only in emergencies. With this, they can actually see into the *Falcon* itself. They scan through the interior, noticing that the *Falcon*'s clock is ticking very slowly. In fact, for every twenty seconds that pass on Tatooine, only one second passes on the *Falcon*. As the stormtroopers scan into the cockpit, they see Han Solo talking very slowly. One, who is adept at lipreading, watches Han patiently, eventually making out two words: "dusting crops." This is truly the most confusing case they've ever investigated. Checking to see that they haven't been observed, they lock up the super-duper scanner and move quickly away.

What the stormtroopers have observed are simply the results of the special theory of relativity. As Han travels closer and closer to the speed of light, they observe the *Falcon*'s length becoming shorter and shorter. The clocks tick slower and slower. And the *Falcon*'s mass becomes greater and greater, requiring more and more energy to accelerate a bit closer to Han's goal. Slowly he gains speed, expending huge quantities of energy. As he approaches the speed of light, he becomes infinitely massive, requiring infinite energy to accelerate that last tiny bit. He'll never make it. Because of this, only objects that have no mass, such as light, can travel at the speed of light. We can't.

Einstein discovered this limitation when he realized that the speed of light will always be measured to be the same, no matter what the observer's speed is relative to the light. This goes against every instinct we have. Let's try another thought experiment. Imagine you are standing on the forest moon of Endor. A speeder approaches you at 50 miles per hour. On it you see Darth Vader and Leia. They are fighting. The handle of the speeder breaks off in Leia's hand, and she throws it at Vader. She has a pretty good arm and can throw the handle at 20 mph. Since she throws the handle in the same direction the speeder is traveling, you measure the handle moving at 70 mph. The handle bounces off Vader's helmet. He's startled, but not hurt. Unhappy with the results, Leia draws a blaster and fires the laser weapon at Vader. The light from the laser leaves the blaster at the speed of light, c. You should therefore measure the light traveling at c + 50 mph. Yet you don't. The light is traveling at only c. How is that possible?

The only way for the velocity of light to remain constant is for your measurements of distance and time to change. Since velocity is calculated by dividing distance traveled by time, these quantities must change just enough so that no matter how fast the speeder is traveling, you will always measure Leia's laser fire traveling at c. In this case, just as with the stormtroopers observing the *Falcon*, the length of the speeder, and similarly the distance traveled by the laser beam, would shrink slightly from your point of view. You would also observe events happening more slowly on the speeder, just as the *Falcon*. You would measure more time passing between Leia's firing of the blaster and Vader's being hit, than Leia herself would measure. These quantities would change just enough so that when you divide distance by time you will get exactly c. Before relativity, we thought time and distance were absolutes. After relativity, the only absolute left is the speed of light.

Einstein's theory of special relativity has frustrated scientists and science fiction fans ever since 1905. If the speed of light forms a barrier, and we can't travel *at* the speed of light, then we can never travel *faster* than the speed of light. If we sent a spaceship to another star, it would not return for years at best. Galactic republics or empires like those shown in the movies could never form. "Star wars" could never occur—or at least not on a timescale that would keep moviegoers interested. So is there no hope that something like *Star Wars* could ever happen, even "a long time ago in a galaxy far, far away"?

SEPARATED AT BIRTH

Queen Amidala gives birth to bouncing baby twins, Luke and Leia. Obi-Wan makes arrangements for the twins to be separated. Leia must be hidden on Alderaan, while Luke will live on Tatooine under Obi-Wan's protection. Obi-Wan decides he will first take Leia to Alderaan, then return and pick up Luke for the trip to Tatooine. Amidala holds the one-year-old twins together for one last time, bids a sad farewell to Leia, and Obi-Wan leaves with the infant. The trip to Alderaan is not far; Obi-Wan decides he can make it without jumping to hyperspace. He travels at a constant speed of .9999 the speed of light. The journey takes one month, during which Obi-Wan wonders why he ever volunteered to change diapers. He entertains himself by braiding Leia's hair into bizarre designs. When he arrives at Alderaan, he discovers the planet is crawling with Imperial forces. He's unable to land safely, and has to turn around and head back to Amidala, spending another month at .9999 the speed of light changing diapers.

He brings Leia from the ship to see her mother. But Amidala appears changed, much older. Obi-Wan decides the separation must have pained her horribly. Then a boy runs into the room and asks to see his sister. Obi-Wan is shocked when Amidala tells him this is Luke. He is now thirteen years old, though his twin Leia is only one year and two months old.

This is the famous "twin paradox" of relativity, and it accurately reflects what would happen in such a situation. Relativity reveals that time is not some absolute cosmic clock ticking the same for everyone. The time that passes for each person is different, depending on where he goes and how he moves. That is because time and space are not two separate qualities, but a single interconnected space-time continuum.

While such effects are minimal in our everyday lives, they become dramatic at speeds close to the speed of light. The good news is that a space traveler can get from one place to another with only a short time passing for him. In our example, Obi-Wan travels to Alderaan in a month. Let's consider how far he traveled.

Before he leaves, Obi-Wan charts his course with Amidala. She notes that Alderaan is 34 trillion miles away, almost six light-years. If he travels very close to the speed of light, it should take him twelve years to make the round trip. But Obi-Wan knows that once he reaches his cruising speed of .9999 c, he will measure the distance between himself and Alderaan as only 482 billion miles, or one-twelfth of a light-year. The distance will have contracted, just like the length of the Falcon in our earlier example. Thus Obi-Wan can make the trip in only two months. According to the time and distance measured by Obi-Wan, he will be traveling at .9999 c. And according to the time and distance measured by Amidala, he will have traveled at .9999 c. Yet in a sense, Obi-Wan has made a trip of twelve light-years while only two months have passed for him, meaning he has traveled at an effective speed of seventy times the speed of light!

The bad news, though, is that much time has passed while he's been away, so that any purpose his trip had is likely obsolete by the time he gets there.

If Star Wars spaceships moved at near-light speeds, this time dilation effect would constantly arise. For example, in *The Empire Strikes Back*, Luke promises Yoda he will return to Dagobah to complete his training. He flies off, has his encounter with Darth Vader at Cloud City, gets a new hand, saves Han from Jabba the Hutt, and then heads back to Dagobah. While Luke may do all this in only a month, for Yoda decades may pass before Luke shows up again. It's enough to try even a Jedi's patience!

Since Luke seems to have spent a fairly uneventful life on Tatooine up until the beginning of *A New Hope*, and Leia seems to have done a fair amount of traveling in her work as a diplomat and a rebel, Leia would certainly be many years younger than Luke when they finally meet.

Such time dilation would make it very difficult for Luke and his rebel friends to destroy the Empire in a timely manner. Han seems to transport the Death Star plans from Tatooine to the moon of Yavin in about a day. Yet years may have passed for everyone else. The Empire could have built a completely new weapon that makes the Death Star look like a squirt gun. Or the Emperor may have died of old age!

WHEN "FASTER THAN A SPEEDING BULLET" JUST AIN'T FAST ENOUGH

So we really don't want to travel at near-light speed to get around. We want to get where we're going before everyone we're going to visit is dead. That means traveling at truly faster-than-light speeds—faster for both the person traveling and the person staying at home. But first things first. Why worry about how to break the speed of light when our current spacecraft can't get anywhere near that speed?

The method we use for space travel today is propulsion, whose workings are explained by Newton's Laws of Motion. According to Newton's second and third laws, the total momentum of an isolated system must remain constant, and every action requires an equal and opposite reaction. As the space shuttle sits on its launching pad, the total momentum of it and its fuel is zero. As the shuttle is launched, the downward momentum of the exhaust gases creates an equal upward momentum in the shuttle. Thus the total momentum remains zero, and the exhaust propels the shuttle upward. In this method, the spacecraft and the fuel are essentially pushing against each other. This makes propulsion particularly suited to space, since in space, there isn't anything much to push against but ourselves.

Some *Star Wars* ships appear to use propulsion, at least under some circumstances. When the *Falcon* comes out of the asteroid field in *The Empire Strikes Back*, it comes under attack by several star destroyers. The rear deflector shields fail, and Han decides to make a frontal assault on one of the star destroyers. As he turns the *Falcon* around and thrusts toward the enemy ship, a white light flares out the back, as if material is being shot out of the ship.

The problem with propulsion, which is obvious if you take a look at the space shuttle on the launching pad, is that you have to carry a lot of fuel, which makes your spacecraft very heavy. And a heavy spacecraft requires yet more fuel to accelerate it. Also, our current chemical engines are not the most efficient at creating high-momentum exhaust, which is what we need to accelerate our ship to high velocities. Momentum is equal to the mass times the velocity, and the velocity of chemical engine exhaust is not terribly high. In a good chemical rocket, the velocity of the exhaust is about 10,000 miles per hour or 2.8 miles per second. That's fifteen-millionths or .000015 times the speed of light. Accelerating our rocket to the speed of the exhaust is not too difficult. To do that, we'll need a mass of fuel only 1.7 times the mass of the ship. That gets our ship to .000015 c. Excited yet?

Since our chemical rockets can't create exhaust with a much higher velocity than this, if we want to increase the momentum of the exhaust, we must increase its mass. To accelerate the ship to twice the speed of the exhaust, we need 6.4 times the mass of the ship in fuel. In this way, we can increase the speed of our ship by shooting out greater and greater masses of exhaust. The problem is that soon the increases in mass become prohibitive. In fact, scientists have shown that to accelerate a space shuttle–sized ship to just .004 times the speed of light, the mass of fuel we'd need is greater than the mass of the entire universe! So we aren't able to get anywhere near the speed of light with current methods of propulsion. What methods might provide better results?

Nuclear fusion is often raised as a possible source of energy for propulsion. There are hints that fusion plays a role in the *Star Wars* universe. When breaking Princess Leia out of the Death

Star's detention center, Han pretends a dangerous "reactor leak" has destroyed all the surveillance devices. Inside the second Death Star, Lando shoots at the "main reactor." The *Star Wars Encyclopedia* even says that "Hyperdrive engines are powered by fusion generators." But could fusion provide that much power?

Going back to our momentum problem, the other way to increase the momentum of the exhaust, besides increasing its mass, is to increase its velocity. As we saw, the velocity of chemical rocket exhaust is pretty low. Fusion can potentially provide exhaust with a higher velocity. In the controlled nuclear fusion envisioned to provide power, hydrogen nuclei or their heavier isotopes fuse to form a helium nucleus. The new nucleus has a bit smaller mass than the sum of the original nuclei. That tiny bit of mass has been converted into a large amount of energy, according to Einstein's equation $E = mc^2$. This famous equation reflects the idea that matter is just a form of energy. Some scientists say matter is frozen energy, or confined or condensed energy. So the fusion process frees a small amount of the energy confined in the hydrogen. That newly freed energy can propel the resulting helium nuclei out the back of the space ship at very high speeds, perhaps one-twentieth the speed of light. Just one ounce of hydrogen fuel can provide as much energy as over seventy thousand gallons of gasoline!

Even though this process is about twenty times more efficient than current chemical methods of propulsion, less than 1 percent of the mass of the original hydrogen is converted into energy. So it's still not terribly efficient. But the higher speed of the exhaust helps a lot. Just as with the chemical rocket, to accelerate the ship to the speed of the exhaust we need 1.7 times the mass of the ship in fuel. But while that amount of fuel brings a chemical rocket to only .000015 c, this amount of fuel brings the fusion rocket to one-twentieth the speed of light, or .05 c. And just as with chemical rockets, if we want to accelerate the ship to twice the speed of the exhaust—in this case one-tenth c—we need 6.4 times the mass of the ship in fuel. Again, though, as we try to accelerate the ship further, the quantities of fuel required rapidly increase. And as we get closer to the speed of light, relativistic effects make the situation even worse.

As our initial thought experiment showed, the stormtroopers

would measure the mass of the *Falcon* increasing more and more as it traveled closer to the speed of light. If we want to accelerate the ship to one-fifth c, four times the velocity of the exhaust, then we need about 57 times the mass of the ship in fuel. The *Falcon* would look more bloated than Jabba the Hutt. To get to one-half c, we'd need over 59,000 times the mass of the ship in fuel. And that's just to accelerate one time! We need an equal amount of fuel to slow down when we reach our destination, and if we're smuggling contraband, we may need to speed up and slow down many times to avoid those pesky "Imperial entanglements."

But what if we didn't need to carry all our fuel with us? An idea introduced by Dr. Robert Bussard in 1960 is to build a ship that can scoop hydrogen out of interstellar space and use it as fuel for fusion. The main problem here is that the density of hydrogen out in space is very low, with an average of only one submicroscopic hydrogen atom swimming in a cube one-half inch on a side. And as we know, we need *lots* of hydrogen to accelerate our ship using fusion. So we'd need a huge scoop miles across to gather up even the tiniest quantities of fuel. This huge scoop would substantially increase the mass of our ship. And if we did make the scoop huge enough to scoop up significant quantities, then the hydrogen in space would provide a resistance to the forward momentum of the ship, like flying into a headwind. So we'd need to work even harder to accelerate. Most scientists agree this isn't a practical method.

Even if we have to carry all our fuel with us, fusion might be useful for small-scale, relatively low-speed transportation and maneuvering, such as between planets within a single solar system or in dogfights above the Death Star. But for long journeys we need something better.

The helium exhaust from fusion can travel at up to one-twentieth the speed of light. Could some other method shoot material out the back of the ship at closer to the speed of light? The best possibility is a method that has long been popular in science fiction, the mixing of matter and antimatter.

Yes, antimatter does exist. The two major theories of physics, quantum mechanics and the theory of relativity, suggest that each particle has a mirror-image antiparticle equivalent, and those antiparticles have actually been found. Antiparticles have the same mass and amount of spin as their particle counterparts, but carry opposite electric charges and a few other opposite characteristics. For example, an antielectron, also called a positron, has the same mass as an electron, but carries a positive charge instead of the electron's negative one. Positrons were discovered in 1932 in cosmic rays, streams of high-energy particles that constantly bombard the Earth. Since then, antiprotons have also been observed.

Antiparticles in and of themselves are no different than what we consider regular particles. If the entire universe were made of antiparticles, life would be pretty much the same. It's when particles meet antiparticles that we run into trouble—or opportunity. The particle and antiparticle can completely annihilate each other, releasing two high-energy photons. Unlike fusion, in which only 1 percent of the mass is converted into energy, here potentially 100 percent of the mass of the particle/antiparticle pair can be transformed into energy, creating an intense burst of radiation. Why does the mixture of matter and antimatter release more energy than nuclear fusion or other processes? If $E = mc^2$, why can't we completely free the energy in any mass?

In any such reaction, certain quantities, such as charge, must be conserved, meaning the totals must be the same before and after the reaction. In most reactions, this prevents mass being entirely converted into energy. But with matter and antimatter, since these quantities are opposite and so carry a net value of zero, the particles can be completely annihilated and all quantities conserved. As the resulting high-energy radiation shoots out the back of the spaceship at the speed of light, the ship would be propelled forward.

You may be wondering how photons of electromagnetic radiation can propel the ship forward. If photons have no mass, then isn't their momentum—the mass times the velocity—zero? No. Since mass and energy are equivalent, the energetic photons do carry momentum, equal to their energy divided by the speed of light. A flashlight aimed out the back of your spaceship could similarly provide momentum, though it would be infinitesimal. The radiation generated by matter/antimatter annihilation would have much higher momentum. In fact, a pound of antimatter fuel can provide as much energy as one hundred pounds of fusion fuel. To accelerate the ship to .99 c, you would need only thirteen times the mass of the ship in fuel. Dr. Miguel Alcubierre, a researcher at the Max Planck Institute for Gravitational Physics in Pottsdam, Germany, agrees that "Matter/antimatter annihilation would be the most efficient way to achieve those speeds."

So why don't we just gather up the antimatter and fire up the engines? Matter fills our galaxy, and antimatter seems to appear only rarely, through radioactive decay, particle collisions, or a few other physical processes. Scientists are now researching why the universe seems made of matter far more than antimatter. While the Big Bang most likely created both matter and antimatter, no antimatter seems to have survived from the beginnings of the universe. Apparently, for some unknown reason, a tiny inequality between the number of particles and antiparticles arose in the first fractions of a second after the Big Bang. For every one billion antiparticles, there were one billion and one particles. The one billion antiparticles and one billion particles quickly annihilated each other, leaving two billion photons of radiation and only one surviving particle on the battlefield. That surviving particle, and those like it, formed all the matter in our universe.

Even though our neighborhood is short on antimatter, we still might be able to use it for fuel. We now have the ability to create antiparticles, and even antiatoms, using particle accelerators. Protons are accelerated and shot at a target, and the collision creates antiprotons. Just one milligram of antimatter, combined with one milligram of matter, can produce as much energy as two tons of rocket fuel. This sounds very impressive; however with current techniques, it will take 200,000 years to create one milligram of antimatter. And the cost will be—well, more wealth than you can imagine. If that isn't enough, it actually takes much more energy to *create* antimatter than we get from annihilating the antimatter with matter. But if those in the *Star Wars* universe have developed a quicker, easier, and cheaper method of creating antimatter, this could be a good method of near-light-speed travel. This may be the method used by the movies' spaceships when they are fighting or traveling short distances.

As we discussed at the beginning of this section, however, near-light-speed travel is just not good enough. To get us quickly across great distances, we need more radical methods of travel. Yet how can we break the speed limit? The answer comes from the very theory that set the limit. Einstein's special theory of relativity was designed to explain what happens only under very specific circumstances; in particular, what happens when we travel at a constant speed and are away from any strong gravitational fields. Einstein chose these conditions because they made the theory easier to formulate. After completing the special theory of relativity, though, Einstein went on to consider what happens outside those specific circumstances, when we are accelerating or decelerating, or when we're traveling near strong gravitational fields. Einstein's general theory of relativity describes what happens in this wider range of circumstances, and so, as Dr. Michio Kaku, professor of theoretical physics at the City University of New York, says, "The general theory trumps the special theory."

The general theory reveals the universe is an even more bizarre place than the special theory implies. While the special theory shows that space and time are interconnected, the general theory shows that mass and energy are also connected to space and time. This opens up a universe of possibilities, ways in which we might use mass or energy to manipulate space to our own ends. Marc Millis, leader of the breakthrough propulsion physics program at NASA, explains that general relativity allows us to "play games with space-time. The trick around special relativity, and the speed limit, is that with general relativity you can warp, fold, distort, or expand space-time." Whether any method of faster-than-light travel might someday work remains speculation, however recent theories suggest there may be ways to break the speed limit—or at least work around it.

Back when the first *Star Wars* movie came out, scientists would not have welcomed such speculation. But Dr. Matt Visser, research associate professor at Washington University in St. Louis, points out that within the last ten years, "The relativity community has started to think about what would be necessary to take something like warp drive or wormholes out of the realm of science fiction." Might we soon be making the jump into hyperspace?

A GALACTIC PIT STOP?

The main problem with propulsion is getting the fuel we need and dragging it around with us. In our galaxy, fuel is not floating conveniently out in space where we can simply pull over for a fill-up. But what if, in "a galaxy far, far away," it were?

If interstellar hydrogen was more plentiful, Dr. Bussard's idea to scoop that hydrogen up to fuel fusion engines might be more practical. Or if antimatter were easily available, Han Solo could have a free source of concentrated energy. Is it possible that another galaxy might have conditions that make interstellar travel much easier than it is in our Milky Way?

Actually, we don't have to leave our galaxy to find plentiful sources of fuel. Near the center of our own galaxy we can find both dense clouds of interstellar hydrogen and plumes of antimatter.

Our galaxy is a huge spinning disk of stars, planets, gas, and dust 130,000 light-years across. Among the 100 billion galaxies in our universe, the Milky Way is fairly average, with about 200 billion stars. The Milky Way is a spiral galaxy, as 70 percent of all galaxies are, with long arms of stars and dust curling out from the galactic core. Our solar system is 25,000 light-years from the core, out on one of the spiral arms. Conditions on the arm are fairly tame. Yet as we move toward the galactic center, conditions become much more dangerous and unpredictable.

The disk of the galaxy bulges in the center with a dense conglomeration of stars several thousand light-years across. As we move through this region toward the galactic core, we find clouds of dust one hundred times more dense than in our neighborhood, pockets of free fuel. Stars whizz about the galactic center at a dizzying one-half the speed of light, packed in a million times more densely. Torrents of high-energy radiation blast out from the core, and intense magnetic fields crisscross the area. At the very center, the dynamo driving all this activity is a black hole or several black holes about two million times more massive than the sun.

These black holes are believed to be the source of antimatter discovered just last year. Dr. William Purcell of Northwestern University found a plume of positrons shooting out from the center of our galaxy. This discovery suggests there may be other antimatter sources in the universe. If a source of antimatter were close to us, and we could develop a safe way of harvesting it, we might use it as a fuel to power our ships. Since we already have safe ways to store the small amounts of antimatter we've created, harvesting and storing the antimatter seems like something we could do. The difficult part would be surviving annihilation if there was a natural source of antimatter nearby. Dr. Visser thinks that could be a problem. "You don't want to find too much around, or that would make life difficult in that vicinity." Curiously, scientists have found no evidence of large quantities of antimatter in the galaxies nearest to us. Yet the jury is still out. Theoretical astrophysicist Michael Burns believes, "It's possible many—maybe all—galaxies have a matter-antimatter source at their cores."

If dense clouds of hydrogen and antimatter exist in the neighborhood of galactic cores, could we theorize that the Republic and Empire of *Star Wars* are closer to the core of "a galaxy far, far away" than we are to our own galactic core?

The core itself would obviously be a very unfriendly place for life, with lethal radiation, ten-billion-degree temperatures, and a black hole whose tidal forces can rip stars to pieces. Dr. Burns agrees. "Within the central parsec of a galaxy, it would be hostile." But how about just beyond that region?

Marc Millis believes denser neighborhoods of stars may offer an advantage: encouraging the inhabitants to develop interstellar travel. "If the nearest star were only one light-year away instead of over four light-years away, as it is for us, that might provoke an intelligent civilization to make the investment necessary to send a probe there and eventually journey there." Dr. Visser agrees. "If a civilization set up shop in a globular cluster or nearer the core of the galaxy, distances between stars would be much smaller, and that would already be a big help."

Yet such a neighborhood might not be the healthiest for humans either. "If you have too many stars too close together," Dr. Visser says, "you run additional risks. A nova could destroy your planet before you can get a civilization formed."

So the best place to start a Republic might be in a neighborhood much like ours. It doesn't look like much, and there's no free fuel, but we have to start somewhere.

"WHEN YOU CAME IN HERE, DIDN'T YOU HAVE A PLAN FOR GETTING OUT?"

When Han Solo takes the *Millennium Falcon* from Tatooine to Alderaan, he makes the "jump to hyperspace," a strange region of stretched-out starlight that seems to allow extremely rapid travel over great distances. This technique must allow Han to "play games" with space-time, warping or distorting it to circumvent the light-speed limit.

One of the weird methods of travel suggested by Einstein's general theory of relativity, which may very well be the one used by Han Solo, is to create a shortcut from one point to another. This tunnel-like shortcut, called a wormhole, allows you to get from your origin to your destination without traveling the entire distance between.

Einstein established that we live in and travel through a flexible four-dimensional space-time, with three dimensions of space and one of time. Anything that folds, expands, or distorts space-time might be made to work for us, to help us travel through it more quickly. How can space-time be distorted? Einstein revealed that any mass distorts it. We call that distortion gravity.

Imagine you are about to skydive out of a plane twelve thousand feet up. When you step out of the plane, the force of gravity is going to pull you toward the massive Earth below. But how does your body know that the Earth is there? Exactly how does the Earth exert this force on your body?

The answer provided by the general theory of relativity is that the mass of the Earth actually distorts space-time, and that distortion pulls your body downward. Since it's pretty impossible for most of us to visualize four-dimensional space-time, instead imagine we live in two dimensions. Space would then be like a huge sheet. Now imagine this sheet is made of a stretchy material like a trampoline. A mass, like a bowling ball, placed on this sheet will create a depression. If you step onto this sheet, you will slide down toward the bowling ball. Similarly, the Earth, placed on this sheet, will create a depression. If you step out of a plane twelve thousand feet away from the Earth, the curvature of the sheet will draw you down toward the Earth's mass, just as you slid down the sheet to the bowling ball. This is exactly what Einstein described in his general theory of relativity.

The heavier the mass, the greater the distortion of space-time. The mass will sink lower and lower into the sheet, creating a deeper and deeper depression, and a steeper and steeper curvature on the sides of the depression.

Now, when I said to imagine space as a sheet, you probably imagined a flat sheet stretching as far as you could see. But as we are learning, the sheet will not remain flat. It will be subject to all sorts of curves and distortions, from every mass in it. So now imagine that this sheet has a large-scale curve to it, like a bedsheet hanging out to dry on a clothesline, half on one side and half on the other. On one side of our sheet, a huge mass creates a deep depression. Now walk around to the other side of the clothesline. On the comparable point on this side of the sheet is another huge mass, creating another deep depression. These two depressions could then theoretically touch and merge, creating a tunnel from one point on the sheet to the other. Without this tunnel, the shortest distance between these two points is to travel the full distance up to the clothesline and down the other side. But with this tunnel, we now have a shortcut that could theoretically take us very quickly from one point to another a great distance away. Dr. Kaku explains, "When your teacher told you that the shortest distance between two points is a line, that was wrong. The shortest distance is a wormhole."

The tunnel does not exist in regular space, since regular space, in this model, is just the two-dimensional sheet. The wormhole exists in an additional dimension, outside regular space-time as we know it. Such additional dimensions are called hyperspace, and we would travel through the tunnel in hyperspace to get to our destination, just as Han Solo does. Perhaps when Han Solo prepares for the jump to hyperspace, he is actually creating a wormhole from his current location to his destination that will serve as an easy shortcut. Using such a shortcut, he could potentially travel from one star to another in a matter of hours. Dr. Alcubierre believes that in *Star Wars* they are using "something similar to wormholes. They jump from one place to another. Wormholes are portals from one place to another far away."

While we've been visualizing the wormhole as a drainlike hole in two-dimensional space, if we translate this back into the three-dimensional space in which we live, the wormhole would actually look like a sphere. What would Han and the others see as the wormhole opened? Dr. Visser says, "If anything, all they would see is a region of space that looked like a window that opened up to a distant region of the galaxy."

Although wormholes have not yet been observed, they are theoretically possible. Unlike black holes, which suck in matter and light and don't let any out, wormholes can potentially be two-way transportation systems, allowing material to both enter and exit. As Marc Millis says, "A black hole is like falling down a well where there's a definite bottom; a wormhole is more like a tunnel."

There are several problems with wormholes, though. If we did manage to find one, the two "mouths" of the wormhole would most likely not be in convenient locations for us. Having one end open in my backyard and the other in Harrison Ford's backyard is not terribly likely, much as I might wish it.

In the off chance that did arise, though, we'd be in for more trouble. It seems that, even theoretically, all wormholes are unstable, forming—if they form at all—only transiently. The two depressions in the sheet meet and create a tunnel only for an infinitesimal fraction of a second; then the tunnel pinches closed. The depressions become separate before anything, even light, has time to pass through them. Worse yet, anything caught inside at the moment of the pinch-off would be crushed in a way that would make getting strangled by Darth Vader seem pleasant by comparison. "Do they naturally exist for any time span in which we could see one?" Dr. Visser asks himself. "I wish I knew the answer to that."

And we needn't be caught at the pinch-off to be killed. Since the deformity of space-time is so great around the wormhole, tidal gravitational forces would most likely kill any human attempting to traverse the wormhole. We talked about tidal forces in connection with planets. The strong gravitational fields of stars can create tidal forces, in which the near side of the planet feels a significantly greater attraction to the star than the far side. It's essentially the same thing with humans. If you imagine yourself dropping feet first into the wormhole, your feet, closer to the wormhole, would be subject to a greater gravitational attraction than your head, and you'd be pulled into a long, thin string.

The size of naturally occurring wormholes might not be terribly convenient either. "I think they are going to be very small," Dr. Visser says, "much, much smaller than atoms." Finding out what's on the other side of these tiny wormholes isn't going to be easy. "Imagine trying to look through a pinhole smaller than an individual atom," Dr. Visser explains. "Ordinary light has a wavelength much larger than an atom. So we can't shine a light through it. I think that's a bit of bad news there. We may get lucky and find one accidentally that's large." Dr. Alcubierre suggests that instead of searching for a large wormhole, we might take a small one and find a way to enlarge it. Without at least a small one to get us started, though, he admits, "It would be hard to think of how to make one in the first place. We'd have to punch a hole through space. I don't think we'd know how to start."

But physicist Kip Thorne at the California Institute of Technology may have found the place to start. He discovered a solution to Einstein's equations that allows for a wormhole that is free of destructive tidal forces and will not pinch closed. This is called a "traversable" wormhole. The trick is that the tunnel or "throat" of the wormhole can be held open if we put something into the wormhole that pushes gravitationally outward on its walls, preventing them from collapsing inward. Since gravity, as we know it, always draws objects together, any material that pushes objects apart, that has in essence repulsive gravity or antigravity (like the device Boba Fett uses to transport Han's carbonite-frozen form), would have to be unlike anything we know of. In fact, some observers would measure this material as having negative energy, something else we've never seen before. Such material is called "exotic matter."

Does exotic matter exist? Dr. Visser points out that a few years ago, the idea would have been laughable. "We would have said, 'Negative matter? That tells us absolutely we can't do these things.' " Yet quantum mechanics, the theory that describes the universe at the subatomic level, suggests that exotic matter does exist. The existence of such negative energy was first predicted by Dutch physicist Henrik Casimir in 1948. In 1958, the effect he predicted was observed. "That force has been measured," Dr. Alcubierre says. "This thing is real." To understand how exotic matter might exist, we need to take a little detour before we return to wormholes. But this detour is important not only for the formation of wormholes, but for antigravity, the Force, and a whole host of other possibilities. Let's see how.

One of the most fantastic realizations of recent years is that the vacuum of empty space is not really empty. The Heisenberg uncertainty principle, one of the main tenets of quantum mechanics, tells us that tiny energy fluctuations may occur that exist for such infinitesimal time intervals that they're impossible to measure. I'll explain this in more detail in Chapter 5, but basically the issue is this: when we measure extremely small, or quantum, quantities, the simple act of measuring disrupts the system. Thus the amount of information we can gather is limited. And if we can't know exactly what's going on at these very small levels, then all sorts of things can be going on.

What do we care if tiny energy fluctuations exist for such short times? Because sometimes this energy may be great enough to create a particle-antiparticle pair spontaneously out of empty space. The pair annihilate each other almost instantly and disappear back into the vacuum before we're able to detect them. (And you thought the Tooth Fairy was hard to believe.) Such particles are called "virtual particles," since they can never be measured directly. But since we can't detect them, physical laws such as conservation of energy, which would normally not stand for something to be created from nothing, remain satisfied.

Scientists envision space on this tiny scale to be a bubbling foam of quantum fluctuations, in which particles, antiparticles, and even quantum wormholes are continuously popping in and out of existence. On the quantum scale, the fabric of space—of reality—actually breaks down. Physicist Heinz Pagels, author of *The Cosmic Code*, compares space to the ocean. "Over long distances the vacuum appears placid and smooth—like the ocean which appears quite smooth when we fly high above it in a jet airplane. But at the surface of the ocean, close up to it in a small boat, the sea can be high and fluctuating with great waves."

If we can't measure these particles, though, how do we know they exist? Scientists such as Dr. Casimir have predicted the effect these virtual particles would have in various situations, and those effects have been measured. So we know that these undetectable, virtual particles do exist. In fact, some scientists believe that the Big Bang out of which our universe formed was triggered by a quantum fluctuation of some kind.

When these fluctuations occur near a strong curvature in space-time, such as near a wormhole, they may lead to the creation of negative matter, matter that has the repulsive gravitational force we need to hold open a wormhole. Imagine that a particle/antiparticle pair pops into existence near the gravitational depression caused by a wormhole. Normally the pair would annihilate each other and disappear. But what happens if one member of the pair is sucked down by gravity into the wormhole, while the other happens to escape into space? The surviving particle then is not annihilated, and can theoretically be detected. It's no longer a *virtual* particle, but a real particle. This is potentially devastating to physics. Remember, conservation of energy was only satisfied because the particles weren't around long enough to detect. Now it appears we have a particle, a bit of energy, created from nothing.

Physicists have found an interesting way to keep their old laws satisfied. Since the surviving particle carries a certain positive energy, then we must assume the particle that fell into the wormhole had negative energy. That way, the total energy would still be zero. This particle with negative energy then qualifies as the "exotic matter" needed to hold open the wormhole.

Is this exotic matter actually produced around a wormhole? We don't know. We still haven't discovered a single wormhole. Even if it is, the exotic matter created might not be enough to do the job. Or it might kill us as we pass through the wormhole. But at least this theory provides one possible source of exotic matter. Whether we might manufacture large quantities of exotic matter to create artificial wormholes is also unknown. "There's no proof that it can't be done," Dr. Alcubierre says. "It may just be that we haven't figured out yet how to do it. Yet it's very doubtful that we'll be able to produce it in such quantities that we'll ever be able to do anything useful with it." Dr. Visser has calculated that to make a one-meter-diameter wormhole, we would need to convert into negative energy a mass equal to that of Jupiter. "You need about minus one Jupiter mass to do the job. Just manipulating a positive Jupiter mass of energy is already pretty freaky, well beyond our capabilities into the foreseeable future."

Since exotic matter, as far as we know, appears only on the quantum level, it would be very difficult to amass an amount this great. Could we do it? "Possibly, yes," Dr. Visser says. "Probably, in the immediate future, no. An arbitrarily advanced civilization might be able to do these things. We, on the other hand, are not arbitrarily advanced."

The *Star Wars* Republic and Empire, however, are. Dr. Kaku believes that a galactic civilization like that in *Star Wars* would have learned how to access huge quantities of energy. He uses astronomer Nikolai Kardashev's method for categorizing future civilizations. A Type I civilization is planetary. It controls the forces on its planet, manipulating the weather and earthquakes, and gaining energy from them. It has colonies on other planets within the solar system. A Type II civilization is stellar. Solar flares are its energy source, and it has begun to colonize neighboring solar systems. A Type III civilization is galactic. It uses the power of billions of stars, black holes, and supernovae, and can manipulate space-time. While Dr. Kaku believes we are a Type 0 civilization, he finds *Star Wars*'s Republic and Empire to be Type III.

Comparing the energy of a planet to that of a star to that of a galaxy, Dr. Kaku approximates that the energy available to a civilization increases ten billion times over with every step up this scale. "How they'll access this energy," Dr. Kaku admits, "I'm not sure. But they will access it." This gives the Republic a lot of energy to play with. Dr. Visser believes that access to those huge amounts of energy would allow the *Star Wars* civilization to produce exotic matter. "If you think of some of the things going on in *Star Wars*, it's likely they could do this."

Exotic matter will only help keep a wormhole open, though. To travel from Tatooine to Alderaan, Han Solo still needs to create the wormhole in the first place. Armed with exotic matter, might those in "a galaxy far, far away" have come up with some clever way to open traversable wormholes—to "punch holes through space"—at will? It's not yet clear whether this might be possible or not. To create huge depressions and deformations in space-time requires we manipulate huge masses and huge energies. "We've found it's going to be extremely difficult if not downright impossible to build these things," Dr. Visser admits. "They will require huge amounts of energy."

This hasn't stopped physicists from proposing a variety of methods to create wormholes, though all remain unproven and pose many problems. As Dr. Kaku says, "Every physicist has his own design. It's like cave men saying, 'Wouldn't it be nice to have a car?' But they don't have any gas. We can design machines, but we don't have the gas." If one of these designs did work, though, physicists in *Star Wars*'s galactic Republic, according to Dr. Kaku's calculations, should have all the gas they need.

GETTING SOMETHING FROM NOTHING

While quantum fluctuations might help produce stable wormholes for interstellar travel, some scientists wonder if these fluctuations might be put to other uses. "Nothing has a little bit of energy," Dr. Kaku says, "and it's all around us." Earlier, we struggled to find free fuel deposits that we could pick up along our interstellar travels. If space itself is foaming with energy, then, couldn't we harness that energy to fuel a ship? Rather than scooping hydrogen out of space, we would use space itself as fuel. Just as with hydrogen or antimatter, one key issue is how powerful the fuel is. Unfortunately, this vacuum energy, also called the zero-point energy, remains poorly understood, and its magnitude is still a matter of great controversy. While the fluctuations we've been discussing are tiny, they permeate all of the universe, and exactly what they would add up to be remains uncertain.

Quantum physicists calculate a vacuum energy so incredibly huge that it would far overwhelm the energy embodied in all the mass in the universe. In fact, it's ten billion trillion trillion trillion trillion trillion trillion trillion trillion trillion times greater. Marc Millis considers this view: "I've read that there is enough energy in the volume the size of a coffee cup to boil away the Earth's oceans—if true and tangible, this could be quite useful." It could certainly help fill up the gas tanks on the *Falcon*. "The amount of energy is so huge," Dr. Alcubierre says, "if we could access it, it would solve all our problems."

Yet cosmologists believe that the vacuum energy cannot be that great, or we wouldn't be here to have this discussion. Energy, they argue, is equivalent to mass, and mass creates a gravitational attraction. If the vacuum energy was huge, the resulting attractive force would have collapsed our universe into a huge black hole in its first moment of existence. Some cosmologists argue that this vacuum energy may be repulsive rather than attractive. But if that is so, a vacuum energy of the magnitude calculated would have such a strong repulsion that it would have blown the universe apart in its first instants, so far apart that not even atoms would have formed. So, these scientists reason, the vacuum energy must be much smaller, though they haven't yet figured out why. Dr. Steven Weinberg, Nobel prize–winning physicist at the University of Texas, asserts that the vacuum energy in a volume the size of the Earth is no more than the energy we get from a gallon of gasoline. Han might as well go back to scooping up hydrogen atoms.

These days, the majority of scientists feel a small vacuum energy is more likely, but a great deal of uncertainty remains. We still don't know whether these vacuum fluctuations create an attractive force, a repulsive one, or neither. In addition, we don't know if we'll ever be able to access this energy. But some scientists remain hopeful that the vacuum energy can be put to good use and are developing techniques to tap into it. They speculate that we may someday learn how to "engineer" the vacuum, as Dr. T. D. Lee describes it, putting "empty" space to work for us.

Even if the vacuum energy is not terribly large, it may still offer ways to make space travel easier and faster. Dr. Hal Puthoff, Director of the Institute for Advanced Studies at Austin, theorizes that quantum fluctuations may be the cause of inertia. His theory remains speculative and extremely controversial, yet it offers a fascinating possibility that could be very helpful to Han Solo.

Inertia is the quality of a body at rest to stay at rest, and a body in motion to remain in motion. It's what throws you back against your seat when you press down on the gas, and what throws you forward when you press on the brake. This resistance to acceleration is exactly what we're fighting against in space travel. We need to accelerate the *Falcon* to high speeds, and to do this we must apply a force of some kind, through propulsion or some other means. The greater the mass of an object, the more force is necessary to accelerate it, as shown in Newton's classic equation, $F = ma$. This equation reflects what we observe, but doesn't explain it. And in the three hundred years since Newton wrote his laws, we still haven't explained it. What is the underlying cause for the phenomenon? Exactly what causes this resistance?

Most scientists simply accept inertia as an intrinsic property of matter. In the near future, though, we may prove that it's something else entirely. Imagine yourself standing in a subway car that has stopped at Times Square station. The doors close, and the subway moves ahead. As it accelerates, you stumble backward, pushed back by inertia. Yet what exactly is pushing you? Dr. Puthoff and colleagues have theorized that quantum fluctuations create an electromagnetic field, called the zero-point field, that causes a resistance on every particle within an accelerating body. "If you stick your hand in a pool of water and try to accelerate it," Dr. Puthoff explains, "you feel a drag force back on your hand. This is the same. If you're standing on a train that takes off with a jerk and suddenly you end up flat on your back, it's as if a two-hundred-pound person knocked you down. The train accelerated you through the zero-point field, just as your hand accelerated through the water." While water and air produce a drag force even if you travel through them at constant velocity, the zero-point field only produces a drag force if your velocity changes. According to Dr. Puthoff's calculations, which are simplified to make the equations manageable, this electromagnetic drag force, amazingly, is proportional to the acceleration of the body and exactly obeys the equation $F = ma$ established by Newton so long ago.

Many scientists remain skeptical of Dr. Puthoff's theory, though. Dr. Alcubierre finds the calculations "fine," but remains skeptical about the conclusions. He believes Dr. Puthoff has proven that the zero-point field produces something similar to inertia. "It's a big jump to say that it *is* inertia."

If Dr. Puthoff's theory is true, this is a bizarre concept: virtual particles are exerting a force on you! Or to be more precise, the electromagnetic forces of these virtual particles are pushing against the electromagnetic forces in the particles that make up your body.

If we can learn to control these quantum fluctuations, we can potentially reduce or eliminate inertia, which would be in essence like reducing or eliminating the mass of an object—which would make accelerating it to high speeds a whole lot easier. Remember, one of the problems of accelerating to near-light speed was that the ship's mass would be measured to increase dramatically. If we could completely eliminate mass, we could theoretically travel at the speed of light, just like photons.

Han talks about making the "jump to light speed." If the *Falcon* is somehow jumping to light speed, it implies a nearly instantaneous acceleration. The *Falcon* might be traveling along at 50 miles per hour, and then suddenly it's traveling at 186,000 miles per second. Let BMW try to beat that acceleration! The *Falcon* might accomplish this by temporarily eliminating inertia. Just eliminating it for a fraction of a second could allow a rapid, effortless acceleration, after which point inertia could return and the *Falcon* could cruise at a constant, high velocity.

Such a technique would also allow humans to survive such an acceleration, and so solve another problem of space travel. It's no problem for Han to accelerate the *Falcon* from zero to 60 miles per hour in five seconds. Inertia will push him slightly back in his seat. But accelerating from zero to 186,000 miles per second in five seconds will push Han back so forcefully that he'll become a splat on that fine vinyl upholstery. The speed of light is so fast, that to accelerate to it safely would take months!

We measure acceleration in g's, with one g equal to the acceleration caused by Earth's gravity—the acceleration of falling objects on Earth. The reason we measure acceleration in terms of gravity is because the two have the same effect. The equivalence principle states that the gravitational force on an object is equivalent to the inertial force on an object undergoing a comparable acceleration. Just as gravity pushes you down against the Earth, inertia pushes you back against your seat.

Imagine you are standing in an elevator at rest on the ground floor of the Empire State Building. The doors close. As the elevator accelerates, inertia pushes you downward against the floor of the elevator, just as gravity pushes you downward against the surface of the Earth. Imagine now that you are in an elevator out in deep space. If the elevator accelerates upward at exactly 32 feet per second^2, the acceleration caused by Earth's gravity, you would be unable to tell the difference between this inertial force and the gravitational force you'd feel if you were simply standing in a stationary elevator on Earth.

We experience higher or lower g forces when we are rapidly changing speeds or directions. Normal humans can withstand no more than 9 g's, and even that for only a few seconds. When undergoing an acceleration of 9 g's, your body feels nine times heavier than usual, blood rushes to the feet, and the heart can't pump hard enough to bring this heavier blood to the brain. Your vision narrows to a tunnel, then goes black. If the acceleration doesn't decrease, you will pass out and finally die. The Air Force's F-16 can produce more g's than the human body can survive. We're forced to limit the acceleration of planes and spacecraft to a level humans can survive.

If we need to accelerate for extended periods, the level we can withstand is even lower. We can withstand 5 g's for only two minutes, 3 g's for only an hour. For the sake of argument, though, let's try to tough it out at 3 g's for a little longer. For Han to take off from Mos Eisley and accelerate at 3 g's to half the speed of light would take him two and a half months. Hardly the makings of an exciting movie. Even at 9 g's, it would take him nineteen days to reach half the speed of light, though he'd be dead long before the ship reached that speed.

Since *Star Wars* ships are constantly undergoing rapid accelerations and decelerations, they must have found some way to solve this problem. Perhaps they have learned to manipulate inertia as Dr. Puthoff suggests. Of course, the force that makes us stumble back as the subway car accelerates doesn't seem *completely* conquered on the *Falcon*. In *The Empire Strikes Back*, the *Falcon*'s jump to hyperspace throws Artoo across the deck and into the open engine pit. Perhaps some of Han's "special modifications" need a tune-up.

WHY MAKE HAN GO TO JABBA, WHEN JABBA CAN COME TO HAN?

Assuming g forces haven't killed Han during one of his wild take-offs, he still needs a faster-than-light method of getting from Tatooine to Alderaan. A wormhole is one possibility. Another possibility is an ingenious method proposed by Dr. Alcubierre. This method, which he calls "warp drive," was inspired by science fiction. "People in *Star Trek* kept talking about warp drive, the concept that you're warping space. We already have a theory about how space can or cannot be distorted, and that is the general theory of relativity. I thought there should be a way of using those concepts to see how a warp drive would work." And that's exactly what he did.

From our earlier discussion, we know that space-time curves and deforms in the presence of mass or energy. Space-time can actually do even more: it can expand or contract. With the Big Bang, the space-time of our universe expanded out from a single point. This doesn't mean that matter expanded out into empty space; it means that space *itself* expanded, carrying matter with it. It's still expanding today.

Scientists theorize that the shape of the universe may be a sphere in four dimensions, or a hypersphere. Let's avoid trying to think in four dimensions by comparing this to a simpler model. Picture our universe as the spherical surface of a balloon. The balloon has tiny little galaxies drawn all over it. As the balloon is inflated, all galaxies move away from ours, as the space-time between them expands. The greater the distance between objects, the more expanding balloon is between them, and so the faster the distance between them increases. This is exactly what we find when we measure the velocity of various galaxies. Space-time is continuing to expand.

If we're going to use this ability of space to expand or contract to help us travel, we need to know how quickly the expansion can occur. Many scientists now believe that in the first fraction of a second after the Big Bang—the first billionth of a trillionth of a trillionth of a second, if you must know—space-time underwent a violent expansion at speeds greater than the speed of light! Yes, you heard me right. While *within* space-time nothing can travel faster than the speed of light, space-time *itself* can travel, or expand, at any speed. The rules of special relativity do not constrain it. This theory, called inflation, suggests that the observable universe is only an infinitesimal portion of the entire universe, and that the great majority of galaxies are so far away that the light from them hasn't had time, in the fourteen-billion-year lifetime of the universe, to reach us yet. In fact, those distant galaxies may even now be moving away from us faster than the speed of light. Or from their point of view, we may be moving away from them faster than the speed of light.

We need to clarify the speed limit, then. A more precise way of saying it would be that nothing can travel *locally* faster than the speed of light. This is one time when an exception to the rule is a good thing. Within a region in which the expansion of space is negligible, we can't travel at or above the speed of light. However when the expansion of space is a significant factor, anything is possible. Let's go back to our balloons. Imagine a clean, deflated balloon. You draw two dots on the balloon one inch apart. You then attach the balloon to the nozzle of a helium tank and rapidly inflate the balloon to full size. The two dots are now perhaps eight inches apart. They have moved quickly away from each other. Yet each dot, in itself, is not able to move and has not moved.

Dr. Alcubierre suggests that, similarly, one could travel great distances without really moving at all, by warping space. Theoretically, after blasting his way out of Mos Eisley, Han Solo could design a space-time disturbance that would expand the space between the *Millennium Falcon* and Tatooine, making Tatooine recede many light years away, and contract the space between the *Falcon* and Alderaan, bringing Alderaan close to the *Falcon*. The *Falcon* would essentially be pushed away from Tatooine and pulled toward Alderaan. As space expands behind it and contracts in front of it, the *Falcon* would be carried toward its destination, like a surfer riding a wave. How fast would such a "wave" travel? Dr. Alcubierre believes the wave could travel with an "arbitrarily large" speed.

The only such wave that seems to have occurred naturally is the one we're riding now in the expansion of the universe. Since it may have once traveled faster than the speed of light, we can theorize that an artificially created wave could similarly travel faster than light. Outside the region of the disturbance, observers would measure the *Falcon* moving faster than the speed of light, just as observing the balloon you would measure the two dots moving quickly away from each other. Yet within the disturbance, Han would not travel faster than the speed of light because light would also be carried along on the wave. Like the dot on the balloon, he would not perceive himself moving at all. One benefit of this technique is that Han wouldn't suffer from inertia. Dr. Alcubierre confirms, "He wouldn't feel any acceleration, wouldn't be squeezed against the back of his seat." Another benefit is that there would be no time dilation effect, as there is when one is traveling near the speed of light. "You can have breakfast at home," Dr. Puthoff says, "take off, have lunch at Alpha Centauri, and come back at dinner and still be talking to your wife, not your great great grandchildren." And going into warp drive might look a lot like what we see when the *Falcon* makes the jump to hyperspace. "My guess is they would probably see something very similar to that," Dr. Alcubierre says. "In front of the ship, the stars would become long lines, streaks. In back, they wouldn't see anything—just black—because the light of the stars couldn't move fast enough to catch up with them."

Using warp drive, Han could reach Alderaan very quickly without actually going anywhere. Dr. Robert L. Forward, scientist and science fiction writer, calls it a "neat idea. Mathematically and physically solid. Just a little difficult from a practical engineering point of view." So how would Han pull this off? To warp space, he would need to manipulate huge masses and energies, just as we discussed with wormholes. Dr. Forward stresses that "The amount of matter needed in front and behind the ship is enormous." Han would also need the same exotic matter with negative energy necessary to hold open wormholes. Remember that exotic matter creates a repulsive, or antigravitational, effect. In this case, the exotic matter is necessary because Han needs to create a repulsive gravitational force between the *Falcon* and Tatooine, expanding space between them. The amount of negative energy required is similar to that for a wormhole. Dr. Visser calculates that to create a warp bubble one meter across, we'd need about the same amount of exotic matter as for a wormhole one meter across: an amount equivalent to the mass of Jupiter. The theory of inflation provides another indication that exotic matter does exist, or at least that it did exist in the first fraction of a second after the Big Bang. Scientists believe its repulsive force may have been responsible for the brief faster-than-light expansion of the universe at that time.

We have one other problem to conquer in the use of warp drive. If our ship travels faster than the speed of light, it is not able to affect the space in front of it. In essence, the ship zips past before space even "knows" it is there. Thus the contraction of the space in front of the ship cannot arise from the ship itself. We need some outside generators to create the contraction in front of the ship. Dr. Alcubierre explains, "We would need a series of generators of exotic matter along the way, like a highway, that manipulate space for you in a synchronized way." The spaceship would then position itself at the beginning of this highway, and the generators would create the wave distortion of space, on which the spaceship would ride. Rather like a train in a subway tunnel, the ship itself wouldn't produce the energy, but would simply ride along on it. This use of a "highway" set up ahead of time doesn't seem like what we see in *Star Wars*, but perhaps with their advanced technology, they've found an easier way to engineer a warp drive.

The ability to warp space could also be used as a defensive mechanism, to deflect enemy laser fire. The *Falcon* has deflector shields, as do star destroyers. How might they work? As we know, a mass creates a depression in space-time that draws other masses to it. Since mass and energy are different forms of the same thing, a depression in space-time can also draw a beam of energy toward it. For example, when light from a star passes near the sun, the sun's gravitational field will attract the light, slightly bending its course. This creates the opposite effect from what we want to occur. The Falcon's mass will actually slightly attract Imperial laser fire toward it. Not a good thing. The sun bends light by only a fraction of a degree, though, so the effect from Han's ship will be infinitesimal. What we want to do is create the opposite effect, and much more strongly.

The opposite of an attractive depression is a repulsive bulge. If Han can warp space, he can create this anti-gravitational bulge. As the star destroyers fire on the *Falcon*, Han could use exotic matter to create a space-time bulge, so that the laser beams would simply bend away. If you consider, however, that the mass of the sun bends passing light rays by only a tiny fraction of a degree, the energy necessary to bend a laser beam enough to make it miss the ship would again be huge—perhaps equivalent to one hundred thousand times the mass of the sun.

THE MUSIC OF THE SPHERES

A third possible method of faster-than-light travel arises from superstring theory. Before you read further, I must caution you. Superstring theory requires that you adopt a whole new view of the universe. This view is consistent with everything we've discussed so far, but it reveals that underlying the visible world, at the smallest level, is a reality completely unlike what we've believed. If you thought particles popping out of nothing and traveling without moving were hard to believe, have I got a whopper to sell you.

Superstring theory attempts to tie together all the forces we observe in the universe in a grand unified Theory of Everything. To do this, physicists found that they had to adopt a new image of subatomic particles. A particle is not a particle at all, they concluded. Instead, a particle is a resonance created by a tiny vibrating string, rather like a musical note created by a plucked guitar string. These resonances appear to us as particles because the strings are very tiny: a billionth of a trillionth of a trillionth of an inch long. Superstring theorists believe that if we could see down to that quantum level—which we can't—we would find not a particle but one of these tiny strings.

Strings with different resonances or frequencies are equivalent to different subatomic particles, such as the quark, electron, and neutrino. As these particles interact, their frequencies can form harmonies, and those harmonies create the physical forces we observe, dictating how particles interact. The Greek mathematician Pythagoras believed that celestial bodies created music as they moved, perfect harmonies that he called the music of the spheres. Now, twenty-six centuries later, he may be proven right.

Why should a particle be anything like a musical note? A musical note is a sound wave made up of energy, defined by a particular frequency. Remember that scientists consider matter to be confined or condensed energy. Thus viewing matter as an energetic vibration confined to the location of a string makes sense.

In addition to adopting a new view of particles, string theorists found they had to adopt a new view of the universe. As they studied how the strings would behave, they found that the tiny strings could not move in just four dimensions. Although we appear limited to a four-dimensional space-time, the universe must actually have ten or more dimensions. And we have a hard enough time visualizing four! These additional dimensions actually make it much easier for scientists to explain how all the different forces we observe are actually different manifestations of a single, unified force. How do extra dimensions help?

Imagine, for example, that you are lost on the planet of Tatooine. You walk and walk across the desert but are unable to find civilization. Finally you come to a rocky outcropping and you climb it. From the top you can see for a great distance, and the geography of the terrain is apparent. You realize that you have been crossing a dry lake bed, and you see at the far end a road that leads toward Mos Eisley. While you were on the ground, working virtually in two dimensions, the geography was difficult to understand. Yet from the higher perspective of the third dimension, the pattern below becomes clear. Similarly, adding dimensions to our picture of the universe adds clarity and reveals previously hidden patterns. But how can the universe have ten or more dimensions, when we don't experience this many?

Imagine that the universe is a giant toilet paper tube and we live on the outside of it. We inhabitants living on the outside of the tube can travel in two directions, along the length of the tube or around the circumference of the tube. Yet what if the diameter of the tube becomes very, very small? Then we inhabitants will believe we are living in a one-dimensional universe, like a string, in which we can only travel along the length. So if one dimension is curled up very tightly, smaller than our ability to measure, we might not know it exists.

Why would this dimension be curled up, or compactified, in the first place? Scientists still haven't found an answer to that. They theorize that during the Big Bang, six of the ten dimensions curled up, while four expanded, as we've described earlier. According to theory, these dimensions are curled up to the Planck length, as Dr. Kaku describes it, "100 billion billion times smaller than the proton." It is at this quantum level that particles foam up out of nothing, and our ideas of space and time break down.

How could these other dimensions allow Han Solo to travel faster than light? If Han could uncurl a dimension, he might be able to take a shortcut through it. This shortcut would be through higher-dimensional hyperspace, like a wormhole. Once he reached his destination, he could theoretically curl the dimension back up again. Like the wormhole, this method could explain the *Falcon*'s jump to hyperspace, quick travel, and return to normal space.

Whether this is possible—and whether the universe actually has ten dimensions—remains uncertain. Marc Millis is skeptical about "assuming the existence of hidden dimensions we can't see and can't interact with." Since we don't interact with them, he argues, "They won't be useful even if they do exist." If they do exist, the energy necessary to access such incredibly small dimensions is huge. This is the Planck energy, ten thousand trillion trillion electron volts, the energy, according to Dr. Kaku, "at which space-time becomes unstable. If we can master the energy found at the Planck length, we will master all fundamental forces."

Our most powerful particle accelerator can give a particle an energy of one trillion electron volts. We need an energy ten quadrillion times greater than that. Yet if we consider that, according to Dr. Kaku's earlier estimate, *Star Wars*'s galactic Republic and Empire have access to energy more than one hundred thousand quadrillion times greater than we do, they could access these tiny dimensions. They might even consider traveling through hyperspace as simple as dusting crops.

WHICH WAY IS UP?

In addition to traveling quickly across vast distances, *Star Wars* spaceships have another intriguing ability. They are able to generate artificial gravitational fields for the comfort of their passengers.

In *A New Hope,* Han Solo alerts Luke that TIE fighters are pursuing them in their escape from the Death Star. Han and Luke stride across the deck to a ladder. Han goes up the ladder, Luke goes down. At each end of the ladder is a pod with a chair and controls for a quad laser cannon. A hemispherical window bulges out from the hull of the ship, allowing the operator to have a wide field of vision. One of these pods protrudes from the top of the *Millennium Falcon,* with the operator's chair facing upward; the other protrudes from the bottom, with the operator's chair facing downward.

What is interesting here is that the scene starts in an area of the ship where artificial gravity has established a clear up and down. Han and Luke are not floating around, but are held to the deck as if it's the surface of the Earth. The ladder seems to maintain this gravity. But if this gravity existed throughout the ship, we'd expect Han and Luke to have a hard time getting into their seats in the pods. In the upper pod, Han would be climbing up into a chair that's reclined at an awkward angle, as if he's visiting the dentist. In the lower pod, Luke would have even a worse time, trying to get into a chair that faces downward. When he reaches the bottom of the ladder, he would find himself hanging helplessly over the pod. If Luke lets go at the bottom of the ladder, he'll fall down onto the hemispherical window. He could instead step down onto the chair, but he'll be standing on its back. To get into it, he has to somehow work himself up into it from underneath, and even if he manages to get into it, only a tightly fastened seat belt will keep him in.

Instead of these bizarre calisthenics, we see Luke step off the ladder, pivot, and get in his seat, as if the chair is upright. And it is. Gravity has clearly shifted. By the time Luke finishes his descent on the ladder, the ladder no longer points down, as it did when he began his descent. Somewhere along the way, a shift in artificial gravity has transformed the ladder into a horizontal catwalk. You can actually see this if you watch the scene carefully. Luke backs off the catwalk into the pod. He is now standing on the side wall of the pod, but with the different gravity in this area, it feels as if he is standing on the "ground." From his point of view, the pod does not feel as if it's pointing down; instead it seems to be pointing out the side of the ship, and the seat is standing upright beside him. This makes it much easier for Luke to get into his chair and blow himself up some Imperials.

With the possible exception of the small fighters, all *Star Wars* ships appear to have artificial gravity. They are somehow able to reproduce the gravitational field of a planet like Earth on a spaceship. And as we saw on the *Falcon*, different areas of the ship might even have different gravities. While we're accustomed to spaceships with gravity in the movies, in actuality, artificial gravity would be very difficult to produce, and except for the crudest methods, we have little idea of how to go about it.

The simplest method, and the only one currently within our power, is acceleration. As we discussed earlier, gravity and the inertia caused by acceleration are equivalent. If a spaceship accelerates at 1 g, those inside will feel an inertial force pressing them against the back of the ship exactly the same as the gravitational force they would feel pressing them against the Earth. The back of the ship then becomes the "ground" to those on board. The prob-

lem with using acceleration to create gravity is that you are forced
to maintain your acceleration at 1 g. If you needed to speed up
or slow down—or heaven forbid, navigate an asteroid field—your
gravity would be shifting all over the place. You'd want to lay in a
good supply of airsickness bags.

To avoid this problem, scientists envision a spaceship shaped
like a wheel that rotates as it moves through space. Since acceleration is caused not only by slowing down or speeding up but by
changing directions, an object moving in a circle undergoes a constant acceleration. If we envision the spaceship as a giant, rotating, hula hoop–type tube with passengers inside the tube, the
rotation would push them against the outermost part of the tube,
which would serve as their "ground." While rotation could produce a steady gravity of 1 g, the ship would still need to speed
up, slow down, or maneuver, as in the first example, and those
accelerations would alter the gravity in the ship, as above. To
avoid that problem, we need to consider some more theoretical
solutions.

In these quite speculative methods, we could potentially compensate for any acceleration of the ship by altering the strength of
our artificial gravity. The artificial gravity would in effect cancel
out any inertial force by creating an equal force in the opposite
direction.

One possible path toward creating artificial gravity would be
to use the zero-point field generated by quantum fluctuations. We
discussed earlier Dr. Puthoff's hypothesis that this zero-point field
is the cause of inertia. Since inertia and gravity are equivalent, you
might expect that Dr. Puthoff believes the vacuum energy is also
the cause of gravity. Indeed, Dr. Puthoff has theorized that a
body's interaction with the zero-point field creates gravity. While
this hypothesis also remains very controversial, if it is so, and if
we could manipulate the zero-point field, then we could potentially create artificial gravity on our spaceships. Since we would be
manipulating the fundamental cause of both inertia and gravity, it
seems as if we could easily compensate for any acceleration of the
ship in any direction. And if we're really the masters of gravity,
we could eliminate the gravitational attraction between a planet
and our ship, making it much easier to take off.

Most scientists, however, remain skeptical about the connection between the zero-point field and both gravity and inertia. Dr. Alcubierre points out that accepting Dr. Puthoff's theory would mean rejecting general relativity, which offers a different explanation of gravity. "That's one of our nicest theories. Why replace a theory that works with something we don't understand?"

Other possible methods of creating artificial gravity arise as side effects of the various methods of faster-than-light travel we've been discussing. Since both wormholes and warp drive depend on the ability to manipulate gravitational forces, chances are that if the Republic has developed that ability, they also have the ability to create artificial gravity. If they can generate the massive gravitational field of a wormhole, they can likely generate a much weaker depression in space-time localized around a ship. That would provide the gravitational field we need.

Dr. Alcubierre suggests a method that would work like the reverse of his warp drive. Just as the warp drive would theoretically expand and contract regions of space to allow a ship to travel from one point to another, an artificial gravity field could be set up by expanding and contracting regions of space within the ship. "If you could produce an expansion above you and a contraction below you," Dr. Alcubierre explains, "this could push you down to the floor."

While these methods may be within the reach of the Republic, they are far beyond our ability or understanding. Dr. Forward stresses that the only well-understood method of creating artificial gravity on spaceships is to rotate them. Other than that, he advocates the old-fashioned method for generating gravity: mass. He suggests "putting megatons of ultradense matter in a thin film under the floorboards." Let's see Han Solo make a quick getaway from the Death Star with that.

WHICH WAY IS DOWN?

While we see gravity in *Star Wars* spaceships, we see antigravity in an even wider range of vehicles. Luke Skywalker's speeder on Tatooine, the speeder bikes on the moon of Endor, Jabba's huge sail barge, and the battle droids' crescent-moon–shaped STAP all appear to utilize some antigravitational field. They seem able to float effortlessly, resisting a planet's gravitational field. To maintain itself at a certain height in the atmosphere, a speeder would have to exert a force equal and opposite to the gravitational attraction of the planet. To accelerate upward into the atmosphere, the speeder would have to exert an even greater force.

Creating antigravity is basically the same problem as creating gravity, except in reverse. And so it could potentially be created using the same techniques discussed above. "If you can distort the geometry of space," Dr. Alcubierre points out, "you can create antigravity." Developing a technology that can easily manipulate gravitational forces not only provides the ability to create gravity and antigravity, but to travel rapidly across the galaxy. Once one of these problems is solved, they all are.

The one difference is that we do have a material at hand that generates gravity; in fact, all matter generates it. But we don't have a material easily at hand that creates antigravity. Exotic matter, if it can be found or made, would be that material. Since exotic matter, in theory, has negative energy, or negative mass, and so creates the exact opposite gravitational effect as gravity, we could use it to cancel out the weight of the speeder. If we measure the mass of the speeder, and then put into the speeder an equal mass of exotic matter, the mass of the total will be effectively zero. Without any mass, the speeder will not be attracted toward the planet or repulsed away from it; the vehicle will tend to stay at whatever height we leave it. Conventional thrusters could then move the vehicle up or down, forward or back, and side to side.

To counteract the mass of any passengers, we'd need the speeder to either produce more exotic matter, or to provide some upward thrust.

You might think that a vehicle able to resist the gravitational attraction of an entire planet would have to access great energies. But the energy needed to hold your car off the ground isn't all that great; your tires do it every day. Your mechanic raises your car even higher when he puts it on a lift. And you yourself resist the force of gravity when you climb stairs, jump in the air, or hang from a tree branch. The problem is that, unlike electric or magnetic forces, we don't yet understand how to manipulate gravitational ones. And you'll get awfully tired hanging from that tree branch.

HAN'S BOAST

Now that we've explored possible methods by which the *Falcon* might travel through space, let's see if we can make sense of Han's incredible boast, that the *Falcon* made the Kessel Run in less than 12 parsecs. If you don't know why Han's boast is so incredible, it's because a parsec is a unit of distance. A parsec equals 3.258 light-years or 19 trillion miles. So when Han brags that he made the Kessel Run in less than 12 parsecs, he's bragging that he made the Kessel Run in less than 228 trillion miles. That's kind of like saying you ran the 100-yard dash in 100 yards. Yeah, so?

Many loyal fans, desiring to save Han Solo from stupidity, have tried to come up with explanations for this seemingly nonsensical statement. Here are a few I've heard:

(a) distortions of time and space at high speeds make a distance measurement just as revealing as a measurement of time

(b) Han discovered a shorter hyperspace route to Kessel

(c) the Kessel Run is a race requiring different cargoes be delivered to different ships, and the ships are moving away as the race goes on. So making all the deliveries while traveling the shortest distance would be very impressive

(d) Han was testing Obi-Wan and Luke to see how gullible they were, so he'd know how much money to charge these two hicks

Author A. C. Crispin even worked an explanation into her book *The Hutt Gambit*, part two of *The Han Solo Trilogy*. Crispin explains that Kessel is next to a cluster of black holes known as the Maw. Reaching Kessel in "less than 12 parsecs" would imply that Han had flown dangerously close to the strong gravitational fields of the Maw, a feat of great daring. It does seem a bit unreasonable, however, for Han to think Luke and Ben would be impressed by this obscure bit of smuggling lore.

Allow me to throw my own two cents into this attempt to rescue Han from stupidity. Let's take a look at explanation (a). As we know, if Han travels toward Kessel at just below the speed of light, the distance he would measure between himself and Kessel would be less than the distance he'd measure if he was at rest with respect to Kessel. The closer he travels to the speed of light, the more the distance contracts. So if Han is traveling so close to the speed of light that the distance between himself and Kessel appears to be only 12 parsecs, that could be a very impressive sign of the *Falcon's* speed. Unfortunately, even if he's traveling at .999999 c, it will take him almost forty years to travel those 12 parsecs and reach Kessel.

So that explanation doesn't work out so well. Most likely, Han is traveling faster than light anyway, considering the distances involved. So perhaps explanation (b) is the correct one. In that case, perhaps Han is using a wormhole to get to Kessel. If he's able to create a wormhole with each end where he wants it, though, he should get to Kessel in much less than 12 parsecs. Even if he can't create a wormhole that ends exactly beside Kessel, say because of black holes or other obstacles, creating one that ends so far away will be pretty inconvenient, though not as bad as our first scenario. In this case, 12 parsecs is the uncontracted distance someone at rest with respect to Kessel would measure. So once Han comes out of the wormhole, he can accelerate to .999999 c and contract the 12 parsecs to just one-twentieth of a light-year. At that speed, it will take him only 20 days to reach Kessel, which is at least better than 40 years. Yet the fans he left on Kessel at the beginning of the race would probably be dead and buried.

The same would hold true if Han was warping space. He should arrive at his destination while hardly moving at all, and if he needs to move short distances at each end of the warp to get to his exact destination, they need to be much shorter than 12 parsecs for him to finish the run in a timely manner.

Explanation (c) sounds like a bizarre game show that I wouldn't mind watching some time. But I'll have to say the most likely seems to be explanation (d). If Ben and Luke had shown a bit more enthusiasm over Han's performance of the Kessel Run, he might have asked them to pay twice as much.

WHEN IN DOUBT, BLAST!

Their light beams crisscross the battlefield, whether it's on the icy plains of Hoth or in the black vacuum of space. They can be carried in the hand, mounted on a speeder, or built within a space station the size of a small moon. They can stun, kill, or destroy a planet. And they make really cool sounds.

We see many light-beam weapons in *Star Wars*. The Death Star's planet-destroying weapon is said in the *Star Wars Encyclo-*

pedia to be a super-laser. Star destroyers are armed with "turbolasers." Blasters are also said to fire beams of "intense light energy." Might we someday have laser weapons similar to those shown in *Star Wars*?

The word *laser* is an acronym, standing for *Light Amplification by the Stimulated Emission of Radiation*. Let's first look at the "radiation" part of laser. Light is emitted, or radiated, by an object when the atoms and molecules that make up that object release energy. If that electromagnetic energy happens to be of a wavelength visible to the human eye, we see it as light. To understand how and when atoms release energy, we need to consider how an atom is put together, with electrons orbiting a nucleus of protons and neutrons. The electrons can orbit at different distances from the nucleus, depending on how much energy they have. Let's do a thought experiment.

Imagine yourself standing in an elevator on the bottom floor of a skyscraper. For our purposes, let's call this floor zero. You receive a jolt of energy that sends the elevator up to the first floor. As you stand on the first floor and the Earth rotates, you are essentially orbiting the Earth. You have a certain potential energy now, reflected by your height in Earth's gravity. If you jump out the window, that potential energy will be converted to kinetic energy. But let's say that instead of jumping, you receive more energy, which kicks the elevator up another flight. You now have more potential energy than before. Your change in energy is exactly equal to the amount of energy the elevator received.

This is similar to the way an electron functions within an atom. When the atom absorbs energy, as when photons of light fall upon it, the electrons are sent up into higher orbits. Later the electrons spontaneously drop to lower orbits and release this energy. Each particular drop—say from the third floor to the second floor, or from the second floor to the first floor—releases a photon with an energy equal to the change in energy of the electron. Let's go back to you in the elevator. If the elevator dropped suddenly from the third floor to the second floor, you might let out a short yelp. If you dropped from the third floor to the first floor, you'd let out a longer scream. You're releasing different amounts of energy depending on the drop, just like the electron.

The energy released by the electron determines the wavelength of the photon it emits. It might be a wavelength within our visible spectrum, or it might be one outside of it, such as infrared, ultraviolet, or X-ray radiation. Such spontaneous emission, as it's called, is what causes light to be emitted from the sun, from incandescent light bulbs, and from fluorescent lights.

While these drops occur spontaneously, we can also stimulate atoms to release their energy at a certain time. That's what the "stimulated emission" part of laser means. Say we have an atom with an electron in an excited state—on the second floor of the skyscraper. Now we radiate that atom with a bunch of photons of the exact energy that our electron would release by going down to the first floor. If you're the electron, it's sort of like hearing the screams of a bunch of your friends jumping from the second floor to the first floor. It sounds so fun, you want to jump too! The photons can actually stimulate an electron to drop down a floor.

If we have a bunch of such atoms, this can cause a chain reaction, with many electrons throwing caution to the winds, jumping, and releasing photons of equal wavelength.

One way to make the most of this chain reaction is to send the newly released photons back through the atoms again, so they will stimulate the emission of yet more photons. Scientists do this by placing mirrors on either side of the atoms, so the light is reflected back and forth through them, the cries of jumping electrons echoing back and forth, stimulating more and more emission. While one of the mirrors is completely reflective, the other reflects only part of the light and transmits the other part through it. This transmitted portion is what we see as the laser beam.

With this technique, we can stimulate the light to be released when we want it. Even better, the light will be of one uniform wavelength and in phase. This means that the peaks and the troughs of all the waves will be lined up with each other, so instead of interfering with each other and canceling each other out, they will add to each other, making the peaks higher and the troughs lower. This is where the "light amplification" part of laser comes in.

While a laser is basically just light, it is light that can be focused onto a precise spot and can have high, extremely concentrated power. Lasers can produce a steady beam for long periods, or they can produce a very intense beam in short pulses. Such pulses can occur thousands or millions of times per second.

The ability to focus on a tiny, precise spot is what allows us to put huge amounts of information—640 million bytes—onto a CD-ROM, which can then be read by a laser. It also allows lasers to perform delicate microsurgery, vaporizing specific cells while leaving others undamaged. Lecturers use pen-sized laser pointers and guns use laser sights because of their ability to remain tightly focused over distances. The coherence and single wavelength of laser light make it useful for conveying information, as it does when it reads the bar codes off your cereal boxes at the grocery store or carries your long-distance phone calls. Lasers print out our documents, remove tattoos, and shatter kidney stones.

The amplified light of lasers can also be very powerful, particularly in intense pulses. The highest-powered lasers, emitting trillions of watts, give off a pulse only a billionth of a second long. A series of pulses can drill through hard materials like titanium or diamond. To have an effect on the material it strikes, though, the laser light must be absorbed by the material. If it is simply transmitted through the object, it will have little effect. If it's reflected by the object, then the shooter is in for a big surprise—just like Han Solo in the garbage masher. This means the laser must be designed to emit light of a wavelength that will be absorbed by the particular material one wants to target. As the target absorbs this intense light energy, it will begin to heat, melt, and then vaporize.

The military has already developed lasers for a number of applications. A low-powered laser can blind enemy sensors or even blind a person. A higher-powered laser could set a person's clothes on fire or burn him, like a long-range flamethrower. A megawatt laser can burn a hole through a jet up to six miles away—though it needs to maintain contact with the aircraft for one to two seconds.

Two of our most powerful lasers can each generate beams with 2.2 megawatts of energy. Alpha is being developed as a space-based laser. From Earth orbit, it would destroy enemy missiles thousands of miles away, using an infrared laser beam with an energy intensity at its core several times that of the sun's surface.

Our other 2.2-megawatt laser, MIRACL, is testing the ability of ground-based lasers to target objects in space. During a test last year, MIRACL was able to hit a satellite in Earth orbit. MIRACL purposely did not destroy the satellite, since the test was designed merely to show that the laser could target and hit the satellite. But researchers say the laser could just as easily have melted it.

Thus it seems the lasers we have today would be capable of doing many of the things we see in *Star Wars*. We could injure or kill people; we could burn structures or melt holes in walls; we could destroy targeted areas of spaceships, assuming we could keep a beam on them for long enough. The main difference between *Star Wars* lasers and ours is the size.

Both MIRACL and Alpha are huge; the two of them together would probably fill the entire *Millennium Falcon*. Even a less ambitious laser, one that could potentially burn a hole through a speeder bike or kill a person, as blasters can do, would be the size of a truck. That would be kind of hard to fit into a holster. This size problem arises from several factors.

One is heat. Lasers have only 1 to 30 percent efficiency, which means that only this much of their energy emerges in the final beam. Most of the rest of the energy, up to 99 percent, is lost as heat. In a very powerful laser, this heat can become very intense, shattering the laser's mirrors. Elaborate cooling systems, involving fans or liquid coolants, must be devised. These are often both large and heavy.

Another factor keeping lasers large is power. While we can create lasers that emit extremely powerful energies, we need to pump great energies into them to make them work. That energy source takes space. While this might not be a problem on the Death Star or a star destroyer, it would be on the *Falcon* or in a blaster. The most compact high-power generator we currently have for laser weapons is thirty-five feet long, eight feet wide, and eight feet tall. The military is very excited that they can fit the fifteen-megawatt generator into a tractor trailer. "You can put the energy of a nuclear power plant on a light beam," Dr. Kaku says, "but that means you have to carry that nuclear power plant with you. There is no portable power pack, other than a hydrogen bomb." That doesn't sound very healthy, and I still don't think you can fit one in a holster. Yet Dr. Kaku admits that "Maybe a Type III civilization has mastered that power. In one hundred years time, we will probably have some form of nanotechnology, the ability to make machines the size of atoms. In which case, rayguns are not such a farfetched idea."

If we can solve the cooling problem and the power problem, perhaps blasters can be in our future. They won't quite be like what we see in *Star Wars*, though.

In the movies, we see a lot of beams shooting through space from one ship to another. A laser beam in space would be invisible. You would only see it if you were looking at the enemy gun the moment it fired or if you were looking at the target as it hit. It would look like a circle of light on the target. The actual beam of a laser only becomes visible if it passes through a lot of dust. The dust scatters some of the light out in different directions, allowing you to stand off to the side and see the beam.

In the movies we also see lasers knocking space ships back as they're hit. I remember feeling the *Falcon*'s pain in *The Empire Strikes Back* as it's struck by a star destroyer's laser and recoils from the impact. Yet lasers do not carry enough momentum to send their targets reeling. Their energy is in the form of intense heat.

We also need to be aware that lasers are often not as effective within an atmosphere. Fog, rain, or smoke, like dust, can scatter and weaken the beam. We could imagine a scene on Naboo where enemy forces are in a pitched blaster battle with Qui-Gon and Obi-Wan. The rain starts to pour down, and suddenly no one's blasters are effective.

Using very powerful lasers within an atmosphere can cause problems. The air absorbs a tiny fraction of the laser light passing through it. If the beam is extremely powerful, like that of a Walker, that tiny fraction can significantly heat the surrounding air, creating turbulence. Turbulence forms areas of higher and lower air pressure, and as the laser beam passes through these areas, it can bend slightly. So a Walker might aim at an Ewok, but end up hitting another Walker instead. The Empire might want to think twice before using those heavy lasers within a planet's atmosphere.

The effect of a laser on the surrounding air is actually used as an advantage in a recently designed laser. This laser can stun, just as the stormtrooper's weapon near the beginning of *A New Hope* stuns Leia. The laser emits a beam of high-frequency ultraviolet light. The intense beam actually breaks apart the molecules in the air, creating a tunnel of positively and negatively charged ions between the laser and the victim. While the laser beam itself does the victim no harm, the weapon immediately sends an electrical current down the ionized tunnel, and the current zaps the victim. Depending on the current's strength, the weapon can cause disabling muscle contractions or induce a heart attack. The weapon has a range of more than one hundred yards. We're still stuck with the size problem, though; this one is as big as a table.

If lasers can be gentle enough to stun, can they also be powerful enough to destroy a planet as the Death Star's laser does? At least size isn't an issue here, since we have a huge space station to play with. Dr. Burns estimates that to vaporize a planet, we'd need a laser with a billion trillion times the energy of MIRACL. But perhaps we don't need to vaporize it. Dr. Stuart Penn, senior research fellow at South Bank University in London, suggests another way a laser might destroy a planet. "The laser could vaporize a narrow tunnel to the core of the planet. Then heat the core so it expands and melts. I'm not sure the planet would actually explode, but the laser would probably rearrange it." The biggest difficulty in generating a beam powerful enough for either of these options would be in finding a stable lasing material—the material whose electrons are doing all that jumping. Lasing materials can be gases, crystals, or even semiconductors. But in very powerful lasers, these materials are subjected to extreme heat. At these high levels, Dr. Kaku explains, "the gas overheats, the ruby cracks." Yet he believes that if the Empire could overcome this limitation, they could build the Death Star's weapon. Dr. Kaku even seems to admire the Death Star a bit. "The Death Star is very practical. We could even build it ourselves, if we had enough gross national product. We have nuclear weapons that could crack the Earth. You can imagine what a Type III civilization could do. They could build a laser powered by a hydrogen bomb, an X-ray laser. I've got no problems with the Death Star."

I have one, actually. When the Death Star fires, six laser beams are generated around the circumference of a circular depression on the exterior of the space station. The six beams meet at the center of the circle and head down toward the planet as a single, huge beam. What would actually happen, I'm afraid, is the six beams would pass through each other and head off in six different directions, probably all of them missing the planet. If we're lucky, maybe they'd run into some star destroyers.

While the Death Star may have been extremely effective in destroying Alderaan, as weapons go it was a bit crude. Let's consider one that's "an elegant weapon, for a more civilized age."

LIGHTNING BOLT ON A STICK

As prominent a role as the light sabers play in *Star Wars*, these weapons seem to be the one thing scientists have no clue how to make. And perhaps that's appropriate, since these weapons channel the Force through them, a mystical energy field that seems to defy every scientific law around (but that's the next chapter).

When I first saw *A New Hope* as a seventeen-year-old, I thought the light sabers were lasers. Yet lasers, as we discussed above, are beams that will continue in a straight line unless they are absorbed, reflected, bent, or scattered by some substance. The light sabers, instead, just stop. Also, the laser beams wouldn't be visible unless there was a lot of dust in the air. And two laser beams would pass right through each other, rather like two flashlight beams.

A much better candidate to create a light saber is plasma. A plasma is a gas that's been heated to extremely high temperatures. Let's go back to our electrons in elevators. As a gas is heated, the atoms of gas move faster and faster, gaining energy. The electrons get more and more shots of energy, moving up higher and higher in their skyscrapers until they at last blow right out through the roof. Such an electron no longer orbits its atom's nucleus but is free and independent. Before, the atom, with its negatively charged electrons and positively charged nucleus, was electrically neutral. Now, although the net charge of the plasma remains zero, the negatively and positively charged particles, called ions, are free to move and act separately. Since electrically charged particles generate electric and magnetic fields, a plasma will act much differently than a regular gas.

In fact, a plasma acts so differently that some scientists consider it a fourth state of matter, in addition to a solid, liquid, and gas. Through their electromagnetic forces, the ions can affect each other's behavior from large distances, so that movement at one end often causes movement at the other. Dr. David Bohm, a protégé of Einstein's who studied plasmas extensively, concluded that plasmas do not behave as individual particles but as a collective, organized whole. In fact, he often had the impression that plasmas acted as if they were alive. This makes a plasma seem all the more appropriate as the main constituent of the light saber.

Yet as bizarre as plasmas are, they are all around us. All stars are made of plasma, including our sun. The outer layer of our atmosphere is a plasma. Plasma can be found in lightning bolts, and we use plasma in the tubes of fluorescent lights. Since the sun, lightning, and fluorescent lights all emit light, it's not hard to imagine that a beam of plasma could produce the glow of a light saber. Light is emitted by a plasma when an electron decides to recombine with a nucleus, dropping back down through the hole in the roof of the skyscraper and letting out a long scream. In most plasmas, the processes of electrons being freed from nuclei and recombining with nuclei are occurring constantly. Fluorescent lights stimulate these processes to create a glow discharge by placing a plasma between two electrodes. The color of the glow depends on the composition and temperature of the plasma.

In Chapter 1, we discussed the activity of similar charged particles, or ions, in space within the Van Allen Radiation Belts. Earth's magnetic field traps ions and causes them to spiral around the magnetic field lines. These magnetic field lines draw closer and closer together near the Earth's poles, where they converge. Since a particle cannot cross a magnetic field line, it must stop, turn around, and spiral back the way it came. The magnetic pole thus serves as a "magnetic mirror." With a magnetic mirror at each pole, the particles are trapped within a certain region, forced to bounce back and forth between the mirrors.

Similarly, physicists use magnetic fields, magnetic mirrors, and electric fields to contain plasmas within a particular region. These fields must exert an inward pressure equal to the outward pressure of the plasma. They can hold a plasma in a cylindrical shape, and they can even control the width of the cylinder by manipulating the strength of the fields. We could thus imagine a long thin cylinder, rather like a light saber, made out of plasma. Yet we have three major problems.

First, as with the laser, we lack a method for creating a clear cutoff for the length of the cylinder. Plasma cylinders are of limited length, since scientists can produce only so much plasma, and they can set up the necessary magnetic fields only within a limited region. Yet we have no method for "capping" the end of a cylinder as seems to be the case with the light saber. Plasma simply leaks out the ends of the cylinder, where the fields fail to trap it. Since the plasma of a light saber must have extremely high energy to cut through metal and skin, it would leak out very quickly. Marc Millis admits, "Light sabers really twist my brain over, how they terminate at a certain distance." Even if we theorize that the light saber's handle continually pumps out more plasma into the cylinder, we should see a diffuse glow at each end as material jets out.

To prevent leakage—which would quickly vaporize the skin off Luke's hand—we need to create a cylinder that has no ends. Scientists do this by curling the cylinder around in a circle, creating a donut-shaped torus. Millis imagines a light saber of this shape would be more possible. "You could have a torus of hot plasma, a donut shape. A stick would go through the hole of the donut and hold the donut. It would work kind of like a mace." We'll have to wait for the next movie for that one. Dr. John Schilling, research engineer at SPARTA, Inc., has another idea for a shape that closes in on itself. He imagines a configuration where the magnetic and electric fields force the plasma to travel up in a very narrow stream in the center of the light saber and come back down along the outside of the saber, creating a tall, skinny plasma "fountain." This could potentially look like the light saber we see in the movies. "Actually implementing such a system," Dr. Schilling says, "especially to the extent of getting the nice, neat cutoff at the tip, would be extremely tough, but not outside the realm of possibility."

Our second problem is familiar from our discussion of the blasters: space. The plasma and all its confining fields are generated from a tiny cylinder about the size of a flashlight. In that space we need a very strong power supply, a cooling system, and plasma sources. If we could create something like a light saber today, all the required systems would fill a building. "The light sabers command a lot of energy in a small amount of space," Millis says. "It has the energy to vaporize a hand off, and in the space of two D-cell batteries. Imagine the Energizer bunny on those."

Third, even if we could generate plasma and an extremely strong magnetic field from the light saber handle, the strength of the magnetic field would decrease quickly as the plasma moves away from the handle. So our leakage problem would become even worse. Placing the handle in the middle with a shorter plasma beam coming out each end of the handle would offer some help with this problem. Yet the two blades on Darth Maul's saber are each as long as a single-bladed saber, so he would suffer the same problem twice over.

If we could solve those problems, a plasma beam could potentially behave as the light sabers do. The confining electromagnetic fields of each beam could repel each other, preventing one beam from cutting through another. But the beams could in theory cut through metal, bone, and other obstacles. Dr. Burns estimates that such a plasma would have to be ten million times more dense than any we've created on Earth, and be ten times hotter, around 200 million degrees. The huge energy density of the plasma would allow it to vaporize its way through just about anything it touches. Unfortunately, heat from the beam would also radiate out to things that don't touch it. This heat would be less intense than the heat from actually touching the beam, just as sitting in the sun heats you less intensely than touching the sun. Yet the plasma is so hot that it would burn anything close by, like Luke's hands, arms, and face. Dr. Schilling says that anything sufficiently close to the saber "would get an industrial-strength case of sunburn. At the least the saber would be painfully hot, and maybe enough to severely burn nearby objects." In that case, perhaps Luke would prefer to mail the saber to Vader and activate it by remote control.

While light sabers, blasters, and everyday trips through hyperspace remain far-off dreams, scientists are beginning to see ways in which these dreams might someday become reality. "There are possibilities," Dr. Alcubierre says. "And I would love to be able to visit other stars." Theoretically, rapid space travel is possible. The trick is in translating theory into reality.

We face a harder challenge with the topic for our final chapter. A theory that allows you to communicate telepathically with other people, move objects with your mind, see events at great distances or even into the future, will not be found in any standard science textbook. Yet the Force raises questions about the nature of our universe that science has been struggling to answer for as long as science has existed. And perhaps, in searching for some answers, we can shed light on this mystical power that underlies all the *Star Wars* films.

5

THE FORCE

Kid, I've flown from one side of this galaxy to the other. I've seen a lot of strange stuff. But I've never seen anything to make me believe there's one all-powerful force controlling everything. There's no mysterious energy field controls my destiny.

—Han Solo, *A New Hope*

It is in every rock, every tree. It is in the air around you, the book before you, the planet beneath you. Your body is not a separate object, but part of an interconnected universal whole. Everything is a part of you, and you are a part of everything. It is always there, like a sound so omnipresent you lose awareness of it. Yet at times, when you are at peace, you can sense the surge of its energy through you. You follow the flow, opening yourself to feelings and information far beyond what your body can tell you. You even, at times, control it, directing its energy toward your own ends.

It can levitate objects. It can transmit thoughts. It can influence the weak-minded. It can reveal visions of the past and the future. It bestows on those in tune with it a sixth sense and life after death. It has a dark side and a light side. It is the Force, the heart and soul of George Lucas's *Star Wars* universe. It is also the most fantastical and least scientific element in that universe. Or at least, no scientific explanation for the Force is ever given. The Force seems to arise from a mixture of myth, magic, and religion. So shouldn't science just keep its nose out?

Yet the very purpose of science is to understand what has not

176

yet been understood. Throughout its history, science has been faced with inexplicable phenomena and invisible forces. As our understanding of them grows greater, the questions we face likewise grow greater. And in its quest to understand the great mysteries—as cosmologist Stephen Hawking says, "why it is that we and the universe exist"—science and mysticism often connect and sometimes overlap. It is in these areas that science might offer us some insight into the Force, and into that ultimate question: Could the Force really exist?

MAY THE FORCE BE WITH YOU?

We can approach this question in two different ways, and these two ways actually reflect two different ways scientists have of looking at the universe. In the first way, we must envision the universe as a collection of elementary particles and packets of energy that interact with each other. In this model, all "forces" are simply the effects we observe when packets of energy, called quanta, are exchanged between particles. Electromagnetic forces, for example, are caused by the exchange of photons. In this model, gravity is not caused by the warping of space as we discussed in Chapter 4, but by the exchange of tiny packets of gravitational energy, called gravitons, which attract matter to other matter.

In this view of the universe, if the Force exists, then it must be carried by some particle or quantum of energy. If Luke is to draw energy from a tree and use that energy to levitate Artoo, then one of two possibilities must occur. First, particles carrying that energy could travel from the tree to Luke to Artoo. This is not as strange as it might seem. Such transfers happen all the time, as when photons of light energy from the sun travel to your face, heating it. Second, a sea of particles could remain stationary, yet allow a wave of some kind to propogate from the tree to Luke to Artoo, as a wave in the ocean passes through the water. Yet since none of the particles we know of behave as the Force behaves, these particles must be unlike any we've seen. And to connect all things, they must be everywhere.

COULD A SCIENTIST BE A JEDI?

When Yoda levitates Luke's X-wing out of the Dagobah swamps, Luke is dumbfounded. "I—I don't believe it."

"That," Yoda says, "is why you fail."

If belief is a requirement for a Jedi, could a scientist be a Jedi? Do scientists believe the Force could be real?

Steve Grand, Chief Technology Officer of Cyberlife Technology:
"It's about time the physicists were given something really awkward to mess up their nice reductionist theories! Who knows what kinds of real, emergent phenomena get conjured up by the frenzied imaginations of a complex universe?"

Dr. Ray Hyman, professor emeritus of psychology at the University of Oregon:
"A force that really covers everything obviously unfortunately doesn't explain anything. It's useless. It's the defining criteria of New Age thinking: everything is connected to everything else. If it is, so what? That doesn't explain a thing. One of the problems of being a skeptic is we're seen as party poopers. People who talk about the Force are people looking for fun, adventure. I'd like to believe there's more romance in real science."

Dr. Hal Puthoff, Director of the Institute for Advanced Studies at Austin:
"If you eliminate precognition, you could almost think of explaining the other properties of the Force with physics as we know it."

Dr. Victor Stenger, professor of physics at the University of Hawaii:
"What they're talking about here is definitely a quantum notion. Especially when you bring consciousness into it and control things with your mind. The really difficult thing to explain is the holistic nature of it. It's a continuous field throughout the universe that acts instantaneously throughout the universe."

Dr. Matt Visser, research associate professor at Washington University in St. Louis:
"The Force makes neither good philosophy nor good religion. My biggest difficulty is I see zero experimental evidence that anything like this exists in reality."

Dr. Jessica Utts, professor of statistics at the University of California at Davis: "It sounds like an alternative explanation for the data we've seen. And it's certainly not one that I would rule out."

Dr. Michio Kaku, Henry Semat professor of theoretical physics at the City University of New York:
"From a Type 0 perspective, it's impossible, it's silly, it violates everything we know about physics. It's hocus pocus, mumbo jumbo New Age nonsense. But what if we look at it from a Type III perspective? That's a technology 100,000 years old. Our technology is only 300 years old. As science progresses, it becomes more like magic, as Arthur C. Clarke observed. At Type III, now we're beginning to enter the realm of magic that is actually physics. What could I do if I had a technology 100,000 years more advanced? Perhaps all those things."

Marc Millis, leader of the breakthrough propulsion physics program at NASA:
"It begs not to be explained. Once it's explained, its magic and mystical allure go away."

THE FIFTH ELEMENT

While the Force combines a unique blend of characteristics, the idea of some unique element that fills all space and so connects all things has been around since ancient Greece. The Greeks believed that all space and everything in it was permeated by an invisible material called the ether. The ether was the substance of the heavens. Aristotle called the ether the quintessence or fifth essence, after earth, air, fire, and water. The ether was said to connect us to each other and to the rest of the universe. As the planets moved through the ether, they sent ripples through it that affected people on Earth. This was how the Greeks explained astrology, the belief that the positions of the stars and planets influence events on Earth.

As science became more sophisticated in the 1800s, our view of the ether became more specific. It had to be massless, since it had not been detected. Yet scientists still believed the ether must exist. Just as ocean waves are propagated through water and sound waves are propagated through air, scientists believed light waves, such as those from the stars, must also travel through a medium, the ether. Newton even proposed that the brain might excite vibrations in the ether, giving humans psychic powers.

Then in 1887, scientists A. A. Michelson and Edward Morley did an experiment that proved the ether did not exist. We came to understand that the electromagnetic radiation that comprises light can propagate itself in a vacuum and needs no medium.

Although no ether exists as the ancients thought of it, could some medium of particles transmit an as-yet-undetected force throughout the universe? One possible candidate for the job is the field of virtual particles continually popping in and out of existence on the quantum level. As we discussed in Chapter 4, the vacuum of space is actually foaming with activity. We discussed then the concept of tapping this vacuum energy to power a spaceship. While we have only the roughest ideas about how this energy might be accessed, let's bypass that problem and consider whether the virtual particles in the vacuum might be able to provide sufficient energy for a Jedi to perform feats to amaze and astound his friends.

Could the zero-point energy potentially be tapped to levitate objects such as Artoo? Let's say Artoo weighs about 220 pounds, or 100 kilograms, and let's assume that the gravity of Dagobah is the same as that of Earth. Then to lift Artoo one yard off the ground will require an energy of about ten billion ergs. This actually isn't as much as it sounds. After all, if Artoo truly weighs only 220 pounds, Chewie can easily lift him a yard. The question is, could the zero-point field provide the energy instead?

If you recall our discussion in Chapter 4, estimates of the magnitude of the vacuum energy differ widely, ranging from so tiny as to be worthless as an energy source, to 118 orders of magnitude—a one followed by 118 zeros—times greater than the energy embodied in all the mass in the universe. If the latter were true, it would certainly fulfill Darth Vader's claim that the power of the Death Star is insignificant compared to the power of the Force.

Recent research, however, may have given us our best estimate yet of the magnitude of this energy. If you recall, cosmologists believe the zero-point energy must be small, because as energy, it should have an effect like mass, most likely creating a gravitational attraction that would make the universe collapse inward. And if for some bizarre reason it instead had an antigravitational repulsion, a high zero-point energy would blow the universe apart.

Scientists had measured that our universe was expanding moderately, and that the expansion was gradually slowing. This made perfect sense without having to include the zero-point energy. The Big Bang initiated the expansion of space by imparting it with a huge amount of energy and propelling it outward. Yet the gravitational attraction of all the material in the universe provides a braking action on expansion. Since the measured expansion of the universe could be accounted for in this way, cosmologists concluded that the zero-point energy must be so small that it plays little to no role in the development of the universe.

Yet in a very exciting experiment last year, two different groups of researchers measured the rate of expansion of our universe is not slowing down, as has long been believed, but is actually speeding up. In fact, they estimate it's now expanding 15 percent faster than it did seven billion years ago, when the universe was only half as old.

This is a seemingly inexplicable result. Why would galaxies, which should be gravitationally drawn toward each other, be racing away from each other faster and faster? Why is space itself expanding ever more rapidly? If the expansion of space is indeed accelerating, then some repulsive force must be operating that is stronger than gravity. The vacuum energy could potentially provide that repulsive, or antigravitational, force.

Scientists used the newly measured acceleration to calculate how great the repulsive force would have to be. While they couldn't calculate the absolute value of this zero-point energy, they could calculate the ratio of how big it is in comparison to the mass energy density in the universe. This ratio is relatively easy to calculate since it compares the strength of the antigravitational force to the gravitational one. What scientists found is that rather than being so small as to be insignificant or so great as to be universe shattering, the zero-point energy density now appears to be a little over twice as large as the mass energy density of the universe.

What exactly does this mean to Luke Skywalker, who wants to levitate his droid? Well, if we can calculate the mass energy density, we will know the zero-point value is twice this. Dr. Michael Burns, a theoretical astrophysicist and president of Science, Math, & Engineering, Inc., provides an estimate of the mass energy density of the universe. By estimating the average energy of a star, the number of stars per galaxy, and the number of galaxies in our universe, he arrives at a very rough estimate of the total amount of mass energy in the universe. Dividing this by the estimated volume of the universe gives a mass energy density of one-hundred-millionth of an erg per cubic centimeter. In a volume of space one meter on a side, the zero-point field provides approximately .01 erg. This means that to get the ten billion ergs necessary to lift Artoo, Luke would need to draw the zero-point energy from a volume of one trillion meters cubed, equivalent to a sphere with a radius of six miles. This is quite a large area to accomplish a relatively minor task, and you can imagine how much larger an area Luke would need to draw energy from to raise his X-wing from the swamp. Although Yoda claims the task is no more difficult than raising Artoo, Luke is unable to do it.

Even though it may be necessary to access the energy from a large volume to amass a significant amount, Dr. Kaku believes that the vacuum energy is the "one thing in physics that comes close to the Force. The only energy that can pervade everything is the zero-point energy."

Yet let's return to that pesky question we bypassed earlier. How could Luke access such energy? It's hard to imagine how a human might use it without any machinery to assist. Particles holding energy of various kinds zip by us all the time, yet we're extremely limited in the ways we can tap into this energy. For example, I can inhale oxygen and use it to burn food to fuel my body. I can step into a beam of light and absorb heat from it. But I can't send that heat energy over to my iguana to keep him warm simply by thinking about it. Dr. Hal Puthoff, Director of the Institute for Advanced Studies at Austin, compares the process to a cartoon with one panel blank except for the words, "And then a miracle occurred."

Yet since the technology in *Star Wars* is so advanced, they may very well have come up with a method of accessing the Force that seems incredible to us today. As Arthur C. Clarke said, "Any sufficiently advanced technology is indistinguishable from magic."

The power of the Force certainly seems magical. Dr. Kaku agrees. "Give me 100,000 years, 10 billion times 10 billion times the energy of the Earth, and then let's talk. If you are in a Type III civilization, then manipulating this force may be possible."

ARE YOU ONLINE WITH THE FORCE?

While we can say that an advanced technology may have some method of manipulating its surroundings with the mind, do we have any idea how they might do this? Dr. Kaku stresses that "True telekinesis I don't think exists; it would violate the four forces of physics," but he proposes that technology might provide an enhancement to biological systems. He cites the case of a paralyzed stroke patient who received a brain implant a few months ago. The implant amplifies his brain signals, which are picked up by a coil placed on his head and transmitted to a computer. Dr. Kaku calls it "radio-enhanced telepathy." The computer can't read his thoughts—we're a long way from that—but with training, the patient has learned what thoughts trigger the computer cursor to move. He has thus learned how to move the cursor to particular icons, and so can communicate with others. He might point the cursor to a "Hello" icon, a "Nice talking with you" icon, or an "I'm hungry" icon. In the next stage of development, scientists will hook up the computer so that the patient can direct the cursor to an icon to turn off the lights or turn the channel on the TV. Dr. Kaku concludes, "Even on Earth within one hundred years, you can easily see the day that by thinking you'll be able to move objects via radio. Your thoughts will be converted into radio signals, sent to a computer, and the computer will carry out your wishes. I can foresee a time, given 100,000 years, when we think and control computers all around us."

Toward that end, Austrian biomedical engineer Gert Pfurtscheller is training a computer to recognize various brain wave patterns. This is a difficult task, because the brain is extremely complex, with over 100 billion neurons, and can potentially process 1,000 trillion pieces of information simultaneously, each one creating an electric signal. Brain waves, then, are a complex hodgepodge of multiple signals. Detecting the exact pattern equivalent to "VCR, play *The Empire Strikes Back*" is impossible for us at this point. To simplify the task for himself, Dr. Pfurtscheller is specifically focusing on mu brain waves, which are associated with the intention to move, actual physical movements, or sensations. Other waves are filtered out of the signal. Since we have conscious control over our movements, mu waves seem a more likely contender than other types of waves for human control, and indeed subjects are succeeding at controlling them. Dr. Pfurtscheller has subjects perform movements such as lifting a finger or smiling over and over while the computer records the brain-wave patterns associated with preparing to make the movement. Once the computer has distinguished these different patterns, Dr. Pfurtscheller can connect each one to a different command. For example, a subject might think about smiling, and the associated mu waves might order the computer to turn off the lights. Thus Luke could think about lifting Artoo, and the computer could lift Artoo for him. "When Luke has to learn how to control this power," Dr. Kaku says, "it's like learning to control his brain waves. There could be a chip in his clothing that picks up the signal."

Similarly, Dr. Grant McMillan and colleagues at the Wright-Patterson Alternative Control Technology Laboratory have built a variety of devices that use the brain's electrical impulses to control lamps, TV sets, video games, and even a flight simulator. So you can turn the channel or potentially fly a plane—or an X-wing—with your mind. And so you might levitate an X-wing—by ordering it to take off—simply by thinking about it.

Thus we might easily imagine humans in the *Star Wars* universe controlling machinery with their minds. Dr. Charles Lurio, aerospace engineering consultant, even theorizes that this technology might have originated long ago, and might reproduce naturally within the biology of current humans, so they might not even be aware of it.

While these are fascinating possibilities to consider, they don't seem to reflect what we see in the movies. Luke doesn't fly an X-wing with his mind. *Star Wars* technology seems to be of the exclusively button-punch variety. And many alien races have the ability to use the Force as well, making a forgotten bio-implant of some kind unlikely. In addition, the Force seems inherently nontechnological. Luke destroys the first Death Star by foregoing technology and trusting his feelings. And most important, even if Luke's brain waves are being relayed to a computer, that doesn't solve the problem of how the computer accesses the Force to levitate Artoo. Rather than worrying about how Luke accesses it, we now have to figure out how the computer might access it. Personally, I find it very hard to believe that the Force is with my computer (unless, perhaps, it's the dark side).

Perhaps in some way, then, Luke can access this vacuum energy to levitate Artoo. But could these virtual particles be responsible for the other powers associated with the Force? Let's consider the sixth sense a Jedi seems to have. The Force tells Luke that something cold and evil lurks in the cave on Dagobah. The Force tells Vader that Obi-Wan has arrived on the Death Star. To understand how this might work, we need to look at how the senses gather information.

A particle can travel from a piece of garbage in the Death Star's garbage mashers up through an air vent to the detention level and into Chewbacca's nose, conveying information that causes Chewie to hesitate diving through the vent. Similarly, the Force tells Luke that something cold and evil lurks in the cave on Dagobah. Luke is sensing something, just as Chewie is. But how? As we discussed earlier, some particle must carry the information to Luke, or a wave in a sea of particles must propagate the information. While some unknown ether might be able to do such a thing, the virtual particles cannot. They pop in and out of existence too quickly to be detected. So having a particle travel from the cave to Luke, or from Luke to Leia when he summons her to rescue him from the underside of Cloud City, isn't possible, unfortunately. Dr. Stenger points out another problem with having the zero-point field carry information. "It's just a random fluctuation, so it's incapable of producing signals."

Are there particles that not only pervade space but live long enough to carry information from one point to another? Dr. Lawrence Krauss, Chairman of the Department of Physics at Case Western University and author of *Beyond Star Trek*, suggests another possible source of an ether that may carry the Force: dark matter. Dark matter is simply matter that we can't see, since it neither radiates nor reflects energy. Then how do we know it's there? Astronomers can tell, from measuring the rate at which a galaxy rotates, that it must be far more massive than the visible matter would indicate. If it weren't, its rotation would cause its stars to fly apart. Scientists now estimate that visible matter comprises at most 10 percent of the total mass of the galaxies. Each galaxy appears to be set within a large halo of dark matter. A galaxy may be about a hundred thousand light-years across; its dark matter halo may be as large as 1.2 million light-years across.

Dark matter may be simply ordinary material, such as extremely faint stars, planets, boulders, cold gas, dust, or black holes. Or dark matter may be made up, not of ordinary material, but of some unusual material we've never even observed before. If dark matter is made up of some unknown particles, it may be all around us and simply undetectable. We could even imagine that these particles are emitted by all things—every rock, every tree.

To consider how such particles might behave, let's look at one type of subatomic particle that now seems as if it makes up at least part of the dark matter in the universe: the neutrino. The neutrino isn't undetectable, but it has proven very difficult to detect. Although neutrinos are widespread, they interact with other matter so weakly that they have been compared to ghosts; it took scientists 26 years to find conclusive evidence that they exist. And it wasn't until last year that Japanese scientists discovered that neutrinos, which were previously believed to have no mass, actually have an infinitesimal mass less than one-billionth that of a proton. While their mass is tiny, there are so many neutrinos throughout the universe that they may add up to a significant percentage of the universe's mass. Huge quantities of neutrinos were created in the Big Bang; more pour from the sun every day; a hundred trillion of them pass through you every second. Since they are so tiny and have no electric charge, they can pass through the Earth without leaving a trace.

The problem with considering neutrinos, or dark matter in general, as the medium transmitting the Force, is that dark matter, by its very definition, is dark. We can't see it. We can't see dark matter because it doesn't interact with matter, or interacts only very weakly. And for Luke Skywalker to get energy or information from dark matter, it must interact with him. Imagine a Sand Person hits Luke with the blunt end of his gaffi stick. The stick strikes Luke in the stomach, imparting kinetic energy to him, sending him stumbling backward. The stick gives energy to Luke because it interacts with him. If the stick flew through Luke as if he weren't even there, not interacting with him, then it could impart no energy or information to Luke.

Neutrinos, which interact very weakly with normal matter, would fly through Luke Skywalker without interacting with him at all. Luke is as invisible to them as they are to him. Could training somehow teach Luke Skywalker how to "make" these particles interact with him? It's hard to imagine how, though that could offer a potential solution. Yet some neutrinos carry more energy than intense gamma radiation. If such particles did interact with him as they streamed through him, they'd be more likely to rip his molecules and cells to shreds than help him levitate a droid.

So if we want to find an ether that can carry the Force, the particles need to be around long enough to transmit information from one place to another; they need to interact with us in a safe way so that we can access that information or energy; and they need to be numerous or energetic enough to make the power to destroy a planet look like peanuts. Do we know of any such particles? No, not yet. There's little doubt in most scientists' minds that unknown particles exist, yet if they did have the qualities above, we very likely would have discovered them long ago.

While the zero-point energy from virtual particles might possibly provide an explanation for part of the Force, it doesn't fulfill all the criteria. In fact, we can see now what makes the Force so hard to understand: the fact that it manifests itself in so many different ways.

But I said earlier that it was possible to approach the Force in two different ways. Let's move now to scientists' second way of looking at the universe. Instead of a universe of particles and packets of energy interacting with each other, this second view describes the universe as a web of continuous force fields. In this view, gravity is caused not by an exchange of tiny particles called gravitons, but by gravitational force fields. Similarly, electromagnetic attraction or repulsion is caused by fields.

To visualize the electrical field around a negatively charged electron, imagine lines radiating out from the electron in all directions like spokes on a wheel. The density or concentration of these lines reflects the strength of the electrical field. As we move away from the electron, the density of field lines decreases, and the strength of the electrical force decreases. As we approach the electron, the electrical force increases. In this view of the universe, though, the electron doesn't even exist! Particles are simply manifestations of fields, areas with an extremely high concentration of field lines. Let's see if this model can lead us to further insights about the Force.

YOU WILL FINISH THIS BOOK, AND YOU WILL ENJOY IT

Luke closes his eyes and sees a vision of the future: his friends being tortured on Cloud City. But how can he see the future, when it hasn't even happened yet? A signal from the future must be somehow traveling back in time to Luke. But doesn't that mean the future has already happened, that it is fixed? Dr. Puthoff believes that of all the strange powers of the Force, "the most difficult thing to get a grip on with physics is precognition."

Many scientists resist the idea that time travel might be possible, since it would violate the principle of causality, the idea that a cause must come before its effects. Indeed, Luke's vision violates causality. If Luke hadn't received the information from the future, he would never have left Dagobah before completing his training. The cause of his leaving—the torture of Han and Leia—actually occurs after the effect—his departure. This is like having me drop dead right now at my computer because someone is going to shoot me in five years. It violates our commonsense understanding of the universe and how it works.

Of course, as we've learned, physics often violates our commonsense understanding of the universe. On the quantum level, the Heisenberg uncertainty principle tells us that we can't measure both the position and the velocity of a particle exactly; there is always a small degree of uncertainty. Thus a particle could travel faster than light for a short time, and we wouldn't even know. Many scientists believe that on the quantum level, this does occur.

On a larger scale, the special theory of relativity prohibits time travel, but the general theory allows its possibility. Since space-time can be distorted, a time distortion might be created. One theoretical method of time travel is through a wormhole. Since a wormhole connects two different points in space-time, some scientists believe a wormhole might connect the future with the past. We might imagine a signal passing through such a wormhole, from the future to the present, showing Luke his friends being tortured on Cloud City.

Another possible method involves tachyons, theoretical particles that can travel faster than light. As we know from the special theory of relativity, as the *Millennium Falcon* travels closer and closer to the speed of light, its clock appears to tick more and more slowly. If it could somehow go faster than the speed of light, the theory reveals that its clock would actually begin to run backward. Thus tachyons, which travel faster than light, appear to move backward in time. Let's imagine that after the torture of Han and Leia begins, some tachyons leave Cloud City and head for Dagobah. These tachyons effectively move backward in time, arriving at Dagobah before the torture actually begins. If the Force allows Luke to somehow access the information carried by the tachyons, he might see the future.

Yet time travel presents problems. If one can see the future, does that mean the future is preordained and cannot be changed? In the case of either wormholes or tachyons, it would seem that we are peeking in at something that has "already" occurred and is unchangeable.

Star Wars avoids this conclusion when Yoda explains, "Always in motion is the future." What exactly we would see in that case is unclear. One possible future? Or an event that occurs in all futures?

While physics doesn't have a problem with time travel, our brains do. We have lived our entire lives in a universe where time moves forward and not back, where causes precede effects, and where the future presents vast, unknowable possibilities. But perhaps someday that will all change. In fact, I know it will.

THE FIFTH FORCE

Just as the early Greeks broke the world down into just four different elements, so scientists have found that all observed phenomena in the universe arise from just four underlying forces. If we want to imagine the Force as a force—and the name certainly fits—then either one or more of the four forces need to be responsible for the phenomena we see, or there must be a fifth force that hasn't yet been detected by scientists.

What are the four forces that govern our universe? Two of them work only on tiny, atomic scales. The strong nuclear force holds together positively charged protons and neutral neutrons in the nuclei of atoms. The weak nuclear force controls the radioactive decay of atoms. The other two forces can operate on larger scales. Electromagnetism attracts negatively charged electrons and positively charged protons to each other, creating atoms and molecules and so all matter. Gravity governs all matter, drawing it together, creating the structure of our solar system and universe.

Scientists have long sought to unify the four theories explaining the workings of these four forces into one grand unified field theory. Such a theory would reveal that these four forces are simply different manifestations of one underlying force, and there are some indications that this is so. Quantum field theory is now able to describe three of the four forces—all except gravity, which, although it may not seem so, is the weakest force. Gravity remains difficult to incorporate into a unified field theory. Superstring theory, which we discussed in Chapter 4, attempts to unite all four, though its work is incomplete. Yet even if we could discover that these four forces are simply different expressions of one underlying force, could that one force be *the* Force?

This unified force acts on us at all times. It acts on every particle in every atom in our bodies. And on every particle in every atom in everything else in the universe. You could certainly say that it "surrounds us and penetrates us; it binds the galaxy together." And you would be quite accurate. But could this unified force do what the Force is said to do?

Unfortunately, these forces, which we understand pretty well, can't convince a stormtrooper that you're innocent, allow you to sense the presence of another Jedi, or bring you visions of the future. In addition, accessing these forces, just like accessing the energy of particles, isn't as easy as thinking about it. Remember that in this view of the universe, particles are simply areas of high concentration of field lines. Thus we might generalize and say a tree, made up of particles, is an area of high concentration of field lines. As Luke accesses the energy in these fields to levitate Artoo, he is moving these fields. Thus the particles of the tree, which embody the strongest points of the fields, should move as well. Yet the particles of the tree are not moving. This is one of the strangest qualities of the Force. Imagine, for example, removing the gravitational field from the moon! (In fact, I've got a bridge in Brooklyn whose weak nuclear forces I'd like to sell you.)

Certainly these forces can be tapped and their energy transferred from one place to another. I can think of several ways to use the four forces in a tree to levitate my little droid. I could cut down the tree, make boards out of it, and create a lever to lift my droid. In essence I'm using the electrical forces between the particles in the atoms of wood, which hold the wood together and make it feel hard, to lift the droid. I could also create an atom bomb using the matter in the tree, and the weak nuclear force triggered would create an explosion that would certainly levitate Artoo, right off of Dagobah! Yet in both cases we are manipulating the matter that embodies the concentrated energy of the field. None of the forces we know of could behave the way the Force does.

So the Force would have to be an as-yet-undiscovered fifth force. Is it possible that we've missed another force in the universe?

Most scientists agree that there probably are undiscovered forces in the universe. Dr. Visser finds this reasonable. "It's not at all unlikely that there might be more than the standard four forces around. The fifth one would have to be pretty weak and satisfy some pretty tight constraints." Since we haven't yet detected such a force, scientists believe it must only act over very short, subatomic distances.

Physicists have been searching for a fifth force, and several have even reported discovering one. But those reports, thus far, have been disproven or remain unconfirmed. Such weak, short-range forces, though, would be unable to have the powers we see the Force having, such as transmitting information long distances.

Even if there was such a force, a "disturbance in the Force" could only travel at best at the speed of light. If Alderaan, many light-years away, is blown to bits, the gravitational field lines will shift as this large mass shoots out in all directions. That shift in gravity will propagate out from the source of the disturbance in a wave traveling at the speed of light. The word "disturbance" seems to suggest a similar wave, like that caused when a stone is dropped into a pond. So Obi-Wan, on the *Millennium Falcon*, could not know about the destruction of the planet instantaneously. Yet he does. It seems as if the disturbance in the Force happens everywhere at once. This violates our understanding of the universe at the most fundamental level.

While an unknown force or particles of some kind might offer possible explanations for some of the powers of a Jedi, the Force presents us with a more fundamental scientific problem. The Force suggests a universe quite different than the one we think we're living in. And it seems equally inexplicable to the more advanced scientists in "a galaxy far, far away." The Force is treated as a mystical religion or superstition, not a scientific reality. Which suggests it is something radically different than anything yet discovered. "If the Force is tangible and they don't have a command of it," Marc Millis says, "it must be a completely different phenomenon." Dr. Visser, research associate professor at Washington University in St. Louis, agrees. "I'm not even sure if it would fit into physics or biology."

So where do we go from here?

WHERE JEDI FEAR TO TREAD

In trying to explain any phenomenon, once science's two views of reality have been considered and no answer found, we have to consider the third option: that reality may be different than we believe.

The Force is said to connect all living things, and this connection doesn't seem dependent on space, or even time. Obi-Wan knows instantaneously of the deaths of the inhabitants of Alderaan. Yoda seems able to "watch" Luke growing up on Tatooine from light-years away on Dagobah, when any signal should take years to travel from Tatooine to Dagobah. Luke is even able to see the future.

These events defy our basic understanding of the universe. Scientists have long believed in the principle of local causality, or locality, that an object can only be affected by something adjacent to it. If I have a mile-long line of dominoes set up, and I tip over the first one, the last one can feel no immediate effect. It will not instantaneously fall over. The first one, instead, will tip over the one adjacent to it, which will tip over the next and the next, following the principle of locality. Only after this process has had time to travel from one end to the other will the second-to-last domino finally fall over and tip over the last domino. The speed at which this effect, or this signal, will travel that one mile depends on the nature of the signal, with the maximum speed possible the speed of light. The Force thus violates locality.

Connected to locality is independence. If locality is true, then each particle is independent, influenced only by those things in its immediate vicinity. Since the Force connects all things in some instantaneous way that transcends time and space, it violates independence as well.

The Force clearly allows Jedi to do things that seem impossible, even magical. Fortunately, science provides a theory that allows for bizarre effects that seem magical, a theory that deals with a realm where a particle doesn't always act like a particle, where locality can appear to be violated, and where distant particles sometimes seem connected in a mysterious way: quantum mechanics, the theory that describes how the universe works at the subatomic level. Dr. Victor J. Stenger, professor of physics at the University of Hawaii and author of *The Unconscious Quantum*, says, "What they're talking about here is definitely a quantum notion."

Out of quantum theory has arisen the controversial scientific hypothesis that most closely approaches an overall description of the Force. This hypothesis, the causal interpretation of quantum mechanics, was put forward by physicist David Bohm, a former protégé of Einstein's, in 1952. Dr. Bohm modified and refined his interpretation up until he died a few years ago. His causal interpretation violates the principle of locality, just as the Force does, and suggests that all things in the universe are interconnected and unified on some deeper level of reality.

Dr. Bohm had a lifelong desire to understand and describe all of reality, and he was troubled and intrigued by some of the implications of quantum theory. Most scientists have preferred the Copenhagen interpretation of quantum theory, originated by Danish physicist Niels Bohr, rejecting Dr. Bohm's interpretation. Yet in recent years Bohm's work has been receiving renewed attention and consideration.

You may be wondering how we can have different interpretations of a theory. Shouldn't a theory describe a physical process without any ambiguity? But in the case of quantum theory, interpretation is key. The theory clearly tells us what results we will measure in different experimental situations, and its predictions have been proven accurate. Yet exactly what these results imply about the underlying reality is unclear. Just as we know inertia exists but we're not sure what causes it, so we know quantum effects exist, but we're not sure how they're created.

In the attempt to explain how, the standard Copenhagen interpretation provides a picture of subatomic reality quite different than what we're familiar and comfortable with in our larger-scale existence. Scientists have for the most part accepted these strange implications, but many have done so only with reluctance, and others, like Bohm, offer alternate interpretations of what this theory actually implies about reality. Although quantum theory has been around for more than seventy years, scientists are still arguing about what it means. Dr. Visser believes the debate is growing ever more active. "There's a lot of speculation going around in the foundations of quantum physics."

Einstein was also troubled by quantum theory, particularly by the interconnectedness of subatomic particles that seems to violate locality and independence and create what Einstein called a "spooky action at a distance." With his colleagues, he explored this "spooky action," which may be similar to what we see with the Force. Their argument became known as the Einstein–Podolsky–Rosen paradox, or EPR paradox.

To understand their argument, let's quickly review some of the basics of quantum mechanics. Don't worry. This will only hurt for a couple minutes. It may twist your brain a bit, but to paraphrase my little green friend (no, not my iguana Igmoe), if you want to understand the Force, you must complete your training. So stick with me on this.

According to the Copenhagen interpretation of quantum mechanics, subatomic particles, such as electrons or photons, can behave either like particles or like waves, depending on how we look at them. In fact, all matter has particlelike and wavelike properties, though the wavelike properties of larger objects are too small to be measured. Just as Princess Leia speaks with a British accent when around Imperial officials and with an American accent when with friends, the behavior of these quantum particles depends on the circumstances in which we observe them.

The second element of quantum theory, the Heisenberg uncertainty principle, we discussed briefly in Chapter 4 in connection with vacuum fluctuations. If you recall, this principle tells us that the amount of information we can gather about a very tiny object is limited, since the simple act of measuring disrupts the object. Specifically, we can't measure both the position and the velocity of a subatomic particle with complete accuracy. We can't reach an accuracy greater than that of Planck's constant. While Planck's constant is a tiny, tiny number, meaning we can be quite accurate on human-scale measurements, this tiny number becomes a significant limitation when we attempt to measure quantities on the quantum scale.

How does measuring a particle disrupt it? To measure the position of an object, we have to shoot a photon or other particle at it. The photon will be bounced or reflected back at us, and we can then "see" or measure the object in question. For example, we might shine a flashlight into a dark room to find the position of Jabba the Hutt. Measuring the position of Jabba the Hutt is no problem, since a few photons aren't going to disturb his blubbery frame. But measuring the position of an electron does pose a problem. Since an electron is so small, the photons from the flashlight will disturb it. We may get an accurate measurement of its position, but its velocity will be radically changed, so we can never know what it was. Or using a different technique, we might measure its velocity, but in doing so change its position.

So quantum theory tells us that we can't know exactly where an electron is and what it is doing. But what does that imply about the underlying reality? What is that electron really up to? The Copenhagen interpretation states that since an electron's position is unmeasurable, it is therefore meaningless. But more than that, the interpretation tells us that the particle doesn't *have* an exact position or velocity until we measure it. The uncertainty of our measurement is not just some peculiarity of the measuring process but an intrinsic property of subatomic particles.

How can a particle not have an exact position? This goes back to the first piece of quantum theory, that a particle is actually misnamed. It may act at times like a particle and at times like a wave, but the most accurate way to describe it is as a wave packet, a localized disturbance whose state is described mathematically by something called a wave function. The wave function gives the probabilities that the particle will be in various states or positions. We like to think of a hydrogen atom as an electron orbiting a proton, like a planet orbiting the sun. But the electron's movement is not nearly so neat. According to the Copenhagen interpretation, the electron has different probabilities of being in different positions. If we envision the electron as Princess Leia, we could say, "She might be in the Hoth command center, or she might be near the *Millennium Falcon*." We do not know exactly where she is. And more than that, she is not exactly anywhere. She is in what physicists call "a superposition of states" in which her wave function encompasses both of these places as possibilities.

In addition to being in no particular place, the electron does not follow a clear trajectory governed by the forces acting on it, like a planet around the sun, or Princess Leia, getting mad at Han Solo and stomping away from the *Millennium Falcon* to visit the command center. As Bohm said, "It is assumed that in any particular experiment, the precise result that will be obtained is completely arbitrary in the sense that it has no relationship whatever to anything else that exists in the world or that ever has existed." In essence, the particle's position is not determined by specific causes or forces. This violates the scientific principle of determinism, which says that a particle's position and state are absolutely determined by the forces acting upon it. So it's not only the Force violating beloved scientific principles; the widely accepted Copenhagen interpretation of quantum theory does as well.

What happens when we measure the position of the electron, or Princess Leia? Measuring, or observing, has a powerful effect in quantum theory. This is the third and final piece of quantum theory you need to know. Before we measure its location, the electron is potentially present in many different places at once, but actually present in none. When we measure it, the electron simply appears in one of the places encompassed by its wave function, without any clear cause determining the particular place. We find Leia in the command center, for no particular reason. We have, in the act of measuring, eliminated the possibility that she is near the *Millennium Falcon*. Similarly, we would find the electron in a specific place and eliminate the possibility that it is in any other place. As physicists say, we have "collapsed" the wave function, eliminating other possibilities and localizing the wave packet, making it take on the properties of a particle. As Dr. Nick Herbert, author of *Quantum Reality,* describes it, "Everything we touch turns to matter."

Dr. Stenger compares the wave function to the lottery. Before the winning lottery number is chosen, or observed, you have a .00001 percent chance of winning. You could win or you could lose—though losing is a lot more likely. When the lottery number is chosen, the probability "collapses." Either you win or you lose. Only one of the possibilities remains as actuality. The requirement that we must observe a particle for it to exist in a specific place means that—at the quantum level, anyway—no objective reality exists. A particle must be observed, a subjective process, in order to exist.

Many scientists have rebelled against the idea that no objective reality exists, since objective reality is exactly what science is supposed to describe. This aspect of the Copenhagen interpretation, along with its violation of determinism, has struck many scientists as absurd and unacceptable. Yet in the absence of a completely articulated alternative, the Copenhagen interpretation has become accepted.

You need just one other piece of information before we can discuss the EPR paradox. Many elementary particles, including electrons, have a quality called *spin*, which reflects the particle's angular momentum. These particles don't actually spin the way the Earth spins on its axis—nothing in quantum mechanics is easy to visualize—but we can imagine them that way without getting into too much trouble. Quantum theory allows that electrons can have only two possible spins with equal and opposite angular momenta. This is comparable to imagining that the electron can spin at only one speed, yet it may spin either clockwise or counterclockwise. These options are called *spin up* or *spin down*. The wave function of an electron, then, may show that it has a 50 percent chance of having spin up and a 50 percent chance of having spin down. Just as with the position discussed above, the electron *has* neither spin, like a coin tossed up and frozen in midair has a 50 percent chance of landing with heads up and a 50 percent chance of landing with tails up. When we measure the electron, though, we force it to display one of these two options. If we think of our coin frozen in midair, the equivalent of measuring the electron is unfreezing time and forcing the coin to land, so that we can measure either heads or tails. If we measure, say, spin up, then the wave function no longer has a 50–50 chance of either spin. It now has only a 100 percent chance to have spin up.

Now let's explore the EPR paradox that bothered Einstein so much, in which an event at one point instantaneously affects an event at a far distant point, violating locality and independence as the Force does. Imagine we have two electrons. We can create a system in which the total spin of the two electrons must be zero. This means that if one electron has spin up, the other must have spin down. The two electrons are, in physics terms, *entangled*: the state of one affects the state of the other. Thus the wave function of the two electrons reveals two possibilities: there is a 50 percent chance of electron 1 having spin up and electron 2 having spin down, and a 50 percent chance of electron 1 having spin down and electron 2 having spin up.

Let's now allow the electrons to move far apart from each other, say one light-year. We then measure the spin of electron 1. Before our measurement, it has a 50–50 chance of having each spin. In our measurement, we find it has spin up. Electron 2, which before our measurement had a 50–50 chance of having each spin, now has a 100 percent chance of having spin down. Since electron 1 has spin up, electron 2 must have spin down. The wave function of electron 2 has collapsed. If Wedge was on hand one light-year away, he could now measure the spin of electron 2, and he would find it had spin down. But how can electron 2 know that it must now have spin down? How can it know what spin electron 1 was measured to have? It seems as if the information was communicated from electron 1 to electron 2 instantaneously, faster than the speed of light! Thus the EPR paradox reveals a situation in which locality and independence are violated, just as when Obi-Wan instantaneously senses the death of all those on Alderaan light-years away.

Since Einstein believed nonlocal effects—which he characterized as "spooky action at a distance"—were impossible, he concluded from the EPR paradox that quantum theory must be incomplete. There must be hidden variables and some underlying reality that we don't yet understand that would explain this effect, that would cause the electrons to have specific spins, rather than just probabilities for various spins. Yet in his rejection of the Copenhagen interpretation, Einstein took an unpopular position.

Other scientists drew different conclusions. Dr. Bohr, origina-
tor of the Copenhagen interpretation, maintained that since quantities such as position or spin are meaningless until they are
measured, we can have no knowledge of what is going on between
the two particles before the measurement is made. Therefore we
can draw no conclusions about whether any "spooky action" has
occurred.

While most physicists were shocked by the implications of the
EPR paradox, just as they had been with many of the implications
of quantum theory, they eventually found an explanation they
could live with. Scientists stress that the above scenario doesn't
truly violate the prohibition that no signal can travel faster than
the speed of light, since this situation does not allow us to transmit
any information from one location to another. Although the electrons may be in faster-than-light communication, we cannot use
this phenomenon to send a warning message to Wedge. He will
either find the electron has spin up or down, and we can't control
which. If this explanation strikes you as very weak, join the club.

Dr. Bohm agreed with Einstein that the Copenhagen interpretation was incomplete, and that there must be hidden variables we
don't yet understand determining the spins of the particles. Thus
there is not really a 50–50 chance of each result, but only one
possible result; we simply lack the information to deduce which
result this is without measuring it. This idea, that the results of
measurements are not simply based on probabilities but are determined by causes and forces we don't yet understand, underlies the
causal interpretation of quantum theory, put forward by Bohm. If
Leia is in the command center on Hoth, then it is not due to chance
but because she was previously at the *Millennium Falcon* and Han
sent her stomping angrily toward the command center. The positions, velocities, and spins of particles are completely determined
by their previous state and the forces acting upon them. Bohm's
interpretation preserved determinism, unlike the Copenhagen interpretation.

Yet Bohm believed the EPR paradox revealed that locality could
be violated. Instead of trying to explain the paradox away, he saw
it as a sign that the universe is connected on some underlying
level. In his view, the two electrons are not two separate particles;
they are one single entity or wave function. Going back to our coin

analogy, we might think of the two electrons as the two faces of the coin. If heads is facing up, tails must be facing down. The two faces are entangled.

Might we not all be part of some cosmic wave function, all entangled and interconnected? This is the question Bohm asked, and the answer, if we pose it to Obi-Wan, would clearly be yes. He is entangled with the people of Alderaan; he is entangled with all things. The destruction of Alderaan has an immediate effect on him. If we ask most scientists, though, we'll get a somewhat different answer. The accepted answer is that entanglement occurs only on a microscopic scale, and not on the macroscopic scale of our experience. As Dr. Kaku says, "According to standard quantum mechanics, there is a wall separating us from the microcosm."

Yet these subatomic particles are part of our world, and so any weird behavior they exhibit affects us as well. In fact, we can construct some bizarre situations in which microscopic events become entangled with macroscopic ones. The oddest of them all is Schrödinger's cat paradox, which Dr. Erwin Schrödinger articulated to reveal how ridiculous quantum theory could be.

In our case, we'll call this the Princess Leia paradox. Leia, by the way, is now playing herself, not an electron. The princess is knocked unconscious and placed inside a sealed cell in the detention block of the Death Star. A blaster is aimed at her. The blaster is connected to a radium atom. Radium is radioactive and will eventually decay. Imagine it has a 50 percent chance of decaying within one hour. When it does decay, it will trigger the blaster and Leia will be shot. The wave function of the radium, then, is a superposition of two possible states: the state in which the radium atom has not yet decayed, and the state in which it has already decayed. Similarly, Princess Leia is then also described by a wave function encompassing two possible states: one in which she has been shot and is dead, and one in which she has not been shot and is alive.

The Copenhagen interpretation states that until Luke opens the door to the cell an hour later and observes Leia, she is neither alive nor dead. Instead, she is in a superposition of states, each having a 50 percent possibility. An observation is necessary to collapse the wave function and put her in one state or the other.

Luke must open the door and look inside for Leia to be either dead or alive.

Dr. Erwin Schrödinger believed this paradox revealed a critical weakness in the Copenhagen interpretation. Einstein agreed, pointing out the silliness of quantum theory by saying, "Does the moon exist just because a mouse looks at it?" Again, we are confronted with the Copenhagen interpretation's conclusion that no objective reality exists, and not only on the quantum level, but sometimes also on the larger scale of our own lives.

Scientists have struggled to find some way in which Leia can be either dead or alive without Luke having to open the door to look at her. The majority of scientists evade this requirement for an observing consciousness by saying that a living observer is not required, simply a measuring device of some kind. Others, like Dr. Kaku, believe that each of these possibilities occurs, splitting the universe into two universes. In one universe Leia is alive; in another she is dead. Each time an observation is made and we see a specific outcome, the universe splits into several alternate universes, one in which each possibility occurs, creating an infinity of universes. According to this theory, these universes do not communicate with each other, so there is no way to prove or disprove that they exist. Yet this "many worlds" interpretation still requires us to make measurements or observations in order to, in essence, "create" our reality. There is no objective reality in these interpretations.

The only interpretation that provides for objective reality is Bohm's causal interpretation, in which hidden forces determine whether an electron has spin up or down, whether a radium atom decays or not, and whether Leia is alive or dead. Whether Luke opens the door or not, these hidden forces will have already acted on the radium atom and determined the situation inside the cell. Leia will not be in some superposition of states. Leia will be either alive or dead. And the result will not be determined merely by chance, like the drawing of a lottery ticket. According to Bohm, our inability to determine the cause of these various events does not mean there is no cause.

This idea of hidden forces led Bohm to differ with the Copenhagen interpretation in other ways as well. Bohm theorized that particles are always particles and never waves. There is no wave function that must be collapsed by an observer for a particle to manifest itself. And each particle always has a specific location, momentum, and spin. We might not know these quantities because of our failure to understand the hidden factors or forces controlling them, but they do have specific values. This certainly agrees more with our everyday experience of reality. And Bohm's causal interpretation is consistent with experimental results. "Some of what Bohm did was absolutely beautiful," Dr. Visser says. "He showed that you can build a hidden variables theory and it works and agrees with experimental physics just as well as the Copenhagen or many worlds interpretations."

Bohm's interpretation raises a challenging question, though. What exactly are these hidden forces that control the movement of every particle in the universe? Dr. Bohm posited the existence of a fifth force, a quantum potential force field that pervades all space. The weird effects of quantum mechanics arise because each particle is accompanied and governed by a wave in this quantum potential field. Just as an electron is always accompanied by an electric field, an electron is also accompanied by the quantum potential field. This quantum potential wave is not like the wave function of the Copenhagen interpretation, a mathematical construct that gives probabilities. It's part of a physical force field, like a gravitational field, that affects the course of particles. Since we don't understand the workings of this force, we can't yet calculate the trajectory of particles. But if we someday understand it, then we can.

Bohm theorized that this force does not decrease with distance, unlike other forces, thus helping to explain long-distance effects as in the EPR paradox. In that situation, a single quantum potential wave controls the two entangled particles, no matter how far apart they go. Measuring a property, such as the spin of one particle, instantaneously alters the shape of the wave, affecting the other particle under its control. So the quantum potential force violates both locality and independence, just as the Force does. The fact that the quantum potential field conveys any change instantaneously over all of space suggests that it carries information about the whole universe.

Dr. Bohm compared the relationship between the quantum potential force and a particle with the relationship between a radio wave and a ship set on automatic pilot to be guided by those radio waves. The radio waves are not pushing the ship, but they provide information that directs the movement of the ship. Similar, one could say, to the way the Force is said to control one's actions.

But could this quantum potential force be *the* Force? At the least it has several qualities in common. First, they are both nonlocal. Dr. Stenger confirms the Force is nonlocal, describing it as "holistic. It's a continuous field throughout the universe that acts instantaneously throughout the universe." Second, the quantum potential field is not created by or concentrated around matter, unlike other forces. The quantum potential force guides particles but is not generated by them, just as the radio waves are not generated by the ship. They simply steer the ship. So while we had a hard time imagining how Luke could remove the gravitational force from the Moon and use it to levitate Artoo, we might imagine Luke gathering and directing the quantum potential somehow, sending signals that "steer" Artoo up into the air of Dagobah. Third, because the quantum potential force permeates all space and is not affected by distance, it in essence entangles or interconnects all particles. In Bohm's view, the universe is not a collection of objects, but a web of vibrating interconnected patterns governed by this quantum potential force. This sounds quite like the view of the universe presented by Yoda. Dr. Visser agrees. "If Bohm is right, if this quantum potential exists, then maybe it is the appropriate framework for doing the things the Force allows."

Aside from its basic properties, we know very little about what this quantum potential force might be like. One possibility, according to Bohm, is that this field may be one and the same as the zero-point field we discussed before. Dr. Puthoff says, "I do think maybe they're the same thing in different clothing." If this is so, then we at least have a sense of how strong the quantum potential is and how it's generated.

Is Bohm's causal interpretation the right one, though? Does the quantum potential force really exist? Since both the causal interpretation and the Copenhagen interpretation predict, with only a few exceptions, the same experimental results, those results support both interpretations equally. While Dr. Stenger finds Bohm's work interesting, he feels that all experimental results can be explained satisfactorily by the Copenhagen interpretation. "There are no phenomena that you need this to explain." He obviously hasn't seen Yoda levitate an X-wing out of a swamp. But what about those few exceptions where the causal interpretation predicts different results than the Copenhagen interpretation? Thus far, those differences have proven impossible to measure. After all, how can we check whether Leia is alive or dead before we observe her, unless we observe her?

Most scientists object to Bohm's idea of the quantum potential force on two grounds, the very qualities the quantum potential has in common with the Force. First, unlike the other forces, the quantum potential appears to have no known physical source. Second, unlike the other forces, the quantum potential force violates locality. If we are to go along with the current accepted wisdom, then both the quantum field and the Force are unlikely. Yet if we consider, as an increasing number of scientists are now doing, that the quantum potential offers an alternative explanation of experimental results, then it seems a valid possibility. And by considering how it may operate, we can learn more about what the Force might be like.

From this point on, our discussion of the causal interpretation of quantum theory becomes less scientific and more philosophical. The evidence can take us no farther. All we can do is speculate about what the existence of the quantum potential might imply about the nature of reality. Yet it is in this no-man's-land between science and philosophy that we may find the Force. "In a sense," Dr. Visser says, "the Force seems to be taking the notion of God and religion and philosophy and putting it into a quasi-scientific setting."

Just as Einstein explained that space and time are interconnected as part of a space-time continuum, Dr. Bohm believed that everything is connected as part of a single continuum, that the entire universe is one single, complex entity. Just as superstring theorists believe the four forces may be projections of a single superforce that exists in ten dimensions, Bohm believed that entangled particles may be projections of a single higher-dimensional reality. Earlier, we discussed how two entangled electrons might be thought of as one single entity, like the two faces of one coin. Bohm used a slightly different example that illustrates how additional dimensions might contain the key to the behavior we observe.

As we know, it's very difficult to imagine more dimensions, so let's pretend that we live in two dimensions. Now let's set up our thought experiment. Imagine an aquarium with a fish inside. We put one video camera at the end of the aquarium, and the other on the side of the aquarium. The aquarium and the fish exist in three-dimensional space. But in our two-dimensional world, all we can see are the images on the two flat TV screens displaying what the two cameras are recording. If the fish is facing the end of the aquarium, on one screen we'd see a tall, very thin creature. On the other screen we'd see an equally tall yet very wide figure. Imagine now the fish turns to the side of the aquarium. On the first screen, the thin creature now becomes wide, while on the second screen, the wide creature becomes thin. As we observe this, we realize that the change from thin to wide on one screen is always accompanied by a change from wide to thin on the other screen. Just as in the EPR paradox, one electron having spin up is always associated with the other having spin down. If we didn't know that these two images showed the same fish, we would wonder how these changes were coordinated. They happen instantaneously, no matter how far apart the two screens are, violating locality. The two images seem to communicate through some faster-than-light method.

The truth, though, is that these two images are not communicating instantaneously. They are simply projections of a single three-dimensional reality. Similarly, Bohm believed the two entangled electrons are facets of a single underlying reality that we are not able to perceive in its entirety. This again relates to what we are told about the Force. Yoda suggests that a greater reality exists beyond what we see, explaining, "Luminous beings are we. Not this crude matter."

Bohm believed that underlying the apparently chaotic physical realm, this higher-dimensional reality holds a hidden or enfolded order. Applying this to the quantum realm, he proposed that the hidden order is maintained by the quantum potential force, which controls and connects all things. Bohm used the hologram as an analogy to explain this underlying order. We discussed in Chapter 4 how a laser emits light of a single wavelength that is in phase. A hologram, in a sense, is like a photograph taken with a laser. It's created by splitting the laser beam into two beams. One beam is bounced off the object to be "photographed" and onto a piece of film. The second beam goes directly from the laser to the film, not touching the object, and mixes with the first. The two beams are now out of phase, and as they come together they create an interference pattern on the film.

If you want to see what an interference pattern looks like, fill your kitchen sink with water. Hold your hands over the water about a foot apart and tap your index fingers regularly into the water. You will see a series of circular ripples emanating from each of your fingers. As these ripples meet, they will interfere with each other. If one peak meets with another peak, they will interfere constructively, creating a higher peak than either original. If a peak meets a trough, they will interfere destructively, canceling each other out. The result is a complex pattern of peaks and troughs.

Holographic film holds a similar image. Unlike a regular photographic negative, the film does not have an image of the object photographed. It looks more like a web of rings and ripples, apparently random and chaotic. Yet when laser light is shone through the film, like light being sent through a slide, the three-dimensional image of the original object is projected, just like the image of Princess Leia that Artoo projects. The interference pattern on the film contains encoded or enfolded within it the image of the object that created it. Shining a light on the hologram unfolds the structure to reveal the image of the object.

For Bohm, the hologram embodied the notion that within something that seems random and disordered, like the image on the holographic film, order is hidden or enfolded, like the image of the photographed object. And there's another property of the hologram that reflects Bohm's view of the universe. If you cut a hologram into pieces and illuminate just one piece with a laser bream, you will see not a piece of the photographed object but the entire object. The image will have lost some sharpness, but still the entire image will be there. Thus in each piece of the hologram, the image of the whole is enfolded.

Similarly, Bohm believed that "The whole universe is in some way enfolded in everything and that each thing is enfolded in the whole." This reflects the interconnectedness implied by the Force and suggests how Obi-Wan might instantaneously know of the deaths of those on Alderaan. If a hologram represents the apparent chaos of the physical world, we might then imagine that the Force is the light through the hologram. If you are trained in the Force, you can see through the seemingly chaotic physical world to the true order and nature enfolded in the universe. In fact, this order and nature is enfolded within you, and by becoming calm, at peace, you can unfold that universe within. You are everything—the rock, the tree, the planet, all people, the universe. You observe this order and are a part of it. You are connected to all things.

Thus science, through the causal interpretation of quantum theory, provides us with an image of the universe very much like that the Force suggests, and with a model of a force that shares many of the properties of the Force.

IT'S ALL IN YOUR MIND

The one key difference between the causal interpretation and the Force is control. Jedi are not simply under the control of the Force; they can also control it. Yet Bohm believed everything is under the control of the quantum potential force. The quantum potential is the *cause* determining the position of every particle. A human consciousness is just one piece of the unified universe, and does not have any special power that a rock or a tree would not have. It is as much under the control of the quantum potential as everything else. Dr. Puthoff explains that in Bohm's view, "Everything is cast in concrete." The cost of creating a universe in which everything is connected through a single, controlling force is that we lose the ability to affect this universe.

Dr. Visser finds this ability of Jedi to control the Force with their minds the most difficult trait to understand. "It's supposed to couple to people's minds. The four forces of physics couple to individual atoms and the body. It's not clear at all how you would set up something that would couple to the mind rather than the body."

To find a universe in which we can affect quantum reality with our minds, we must give up the idea that everything is connected and return to the concept of locality. This takes us back to the Copenhagen interpretation. In the Copenhagen interpretation, consciousness plays a special role. It is necessary to collapse the wave function and so to "create" reality. While the vast majority of scientists believe a measuring device can serve the same purpose, some believe a conscious observer is required, the mind playing some special role in this process.

Dr. Henry Stapp at the University of California at Berkeley is one of those who believes consciousness holds a special place in the universe. In his theory, based on the Copenhagen interpretation, a measuring device is not sufficient to collapse the wave function. Let's go back to the sealed cell with Princess Leia and the blaster. The blaster, remember, is set to fire if a radium atom decays, which it has a 50 percent chance of doing within an hour. Leia remains in a superposition of dead and alive states until an observation is made and her wave function collapses. At the end of the hour, instead of sending Luke in to observe Leia, we instead send a droid-cam. The droid-cam enters and records what it finds.

Most scientists would believe this recording or measurement is sufficient to collapse the wave function of the radium particle, and so to put Leia into a state in which she is definitely alive or dead. But Dr. Stapp believes a conscious observer must read the measurements, or watch the recording, in order to collapse the wave function. If this is so, then the recording made by the droid-cam has now become entangled with the radium atom and Princess Leia. Its recording is in a superposition of states: one in which its recording shows Leia has been shot, another in which the recording shows Leia is fine. Similarly, when the droid-cam plays the recording for Luke, he too becomes entangled in this mess. Yet for some reason, he does not go into a superposition of states, one in which he has seen a recording of Leia dead and one in which he has seen a recording of Leia alive. Instead, as his conscious mind observes the recording, the wave function collapses. The introduction of an observing consciousness changes everything. Only then does the recording become one either of Leia dead or Leia alive; only then does Leia actually become either dead or alive; and only then does the radium atom become either decayed or not decayed.

Why does a conscious mind play this special role? Dr. Stapp believes that when the atoms in the brain of the conscious observer become entangled with the atoms being observed, a special opportunity is created: mind and matter can interact. He has even theorized that an additional term is required in the equations of quantum theory, a term that reflects this entanglement of the observer's brain and the observed particles. If such a term is valid in these equations, not only could the human mind collapse the wave function, but it could potentially affect *how* it collapses—whether Leia will be dead or alive. Dr. Robert Jahn, director of the Princeton Engineering Anomalies Research Program, studies physical phenomena that appear to correlate with conscious intention. He believes that his experimental findings require a science that "involves the human participant in the determination of the results. This is obviously a radical departure from conventional physical science."

Before we go any further, I must stress that most scientists firmly believe that the state into which a wave function collapses is merely a matter of probabilities and chance. Measuring or observing constrains the wave packet and so forces it into a particular state, but measuring or observing cannot control into which state it goes. Dr. Stenger believes that if we had such a power, we would have seen its effects long ago. "We have to go by what we know, our established observations about the universe. What we have is a theory that's been around for 93 years and never been found to be violated. That doesn't mean it won't be someday, but based on what we know, we can't make any such assumption."

Those who would like to believe that the mind can affect reality—for example, those who believe in the paranormal—use the Copenhagen interpretation or theories like Dr. Stapp's to help justify their beliefs. Dr. Stenger believes these ideas are very poorly based. "They use the idea in the Copenhagen interpretation of quantum mechanics, that the reality of a body doesn't come into being until you observe it. The place they go wrong is to assume that the kinds of effects that can occur at the quantum level will occur at the macroscopic scale of the human brain. The human brain is a piece of classical machinery. Quantum mechanics has very little to do with the human brain, or the human body as a whole. We need quantum mechanics to understand the chemicals in a rock, but that doesn't mean that rock is conscious."

Yet some believe that theories like Dr. Stapp's indicate that

mind may influence matter on the quantum level. If the collapse
of the wave function is due to some mind/particle interaction,
they argue, then why couldn't mind influence matter? Dr. Puthoff
agrees with this thinking. "I do believe consciousness affects reality."

What led Dr. Stapp to his radical theory? The experiments of
paranormal investigator Helmut Schmidt. Schmidt generated a series of random negative and positive numbers using a random
number generator (RNG), an electronic device that uses radioactive decay to generate a sequence of random numbers. As we discussed with the Princess Leia paradox, the time of radioactive
decay cannot be predicted, so it serves as a good source of random-
ness. The RNG cycles over and over through the possible numbers,
like a person flipping through a deck of cards. At the moment the
radioactive decay occurs, the number currently in the cycle—the
card on top of the deck—is emitted by the machine. This can occur
a few hundred times a second, so a sequence of random numbers
can be quickly generated.

Just as the radium atom in Leia's cell determined whether the
blaster shot her or not, these radioactive atoms determined which
numbers were generated based on whether the atoms decayed or
not. Just as Leia's condition was recorded, the numbers generated
were recorded. If Dr. Stapp's theory is correct and this recording
is not sufficient to collapse the wave function, then the recording
would be in a superposition of states, with many different possible
lists of different numbers. This recording would not collapse into
a single list of numbers until an observer looked at it.

In Schmidt's experiment, no one, not even Schmidt, looked at
the list of numbers for several months. Then the list was scrolled
across an electronic display before a class of martial arts students.
The students were asked to try to mentally influence the display
to show more positive numbers than negative ones. The numbers
revealed were, indeed, more positive than negative, so much so
that the results had less than a one-in-a-thousand chance of occurring at random. Dr. Stapp theorizes that the numbers did not collapse to a single list until observed by the students, and that their
mental states affected how the numbers collapsed, making them
more positive than negative. In essence, the students were influencing the radioactive decay that had occurred months earlier.

If this were possible, we could potentially transmit information in the situation set up in the EPR paradox, information that would travel faster than the speed of light. In that case, we had two electrons that were entangled so that their spins had to be opposite each other. If we measured electron 1 with spin up, we knew electron 2 must have spin down. Electron 2 would somehow instantaneously "know" that electron 1 had been measured with spin up, even if it was a light-year away. The way physicists explain themselves out of this jam is that we can't use this phenomenon to transmit any information, so it doesn't really violate the light-speed limit. If Wedge measures the spin of electron 2, we can't control what he's going to measure. It will simply be the opposite of whatever we measured.

But if we can control what we measure, by mentally influencing electron 1 to have a particular spin, we can send a message to Wedge. We could arrange with Wedge that if he measures the electron to have spin up, he should attack the Empire by land, whereas if he measures spin down, he should attack by sea. We could then, using our mental powers, make our electron have spin up, so that Wedge's will have spin down, signaling him to attack by sea.

Believers in the paranormal view experiments like Schmidt's as evidence that the mind can affect quantum events. And if that is true, they believe these quantum effects could provide the mechanism through which other paranormal powers could operate, powers such as clairvoyance, telepathy, psychokinesis, and precognition. The very powers of the Force.

Yet experiments testing paranormal powers are often not as definitive as they might seem. While Dr. Stenger respects Dr. Stapp's attempt to quantify the role of consciousness in equation form, he finds these experimental results "fishy." Many different factors play a role in the validity of experimental results, and we'll go into these more later. But experiments testing the existence of these powers offer another approach to studying the Force. Whether we can explain it theoretically or not, have the phenomena associated with the Force been observed? Does the Force really exist, right here on Earth?

INVADERS FROM THE FOURTH DIMENSION

I f the Force exists, it may well exist in a greater-dimensional space than we can perceive. Bohm's theory suggests that the Force may exist in additional dimensions. Superstring theory similarly posits that our universe may have ten dimensions. Whether or not we find the Force in these additional dimensions, they themselves may provide us with amazing abilities. In Chapter 4 we discussed how we might travel quickly across the galaxy by uncurling a collapsed dimension. Just as we imagined using a higher dimension to shortcut through space, we might imagine using a higher dimension to gain the powers attributed to the Force.

Carl Gauss, the famous nineteenth-century German mathematician, used a thought experiment to clarify the effects of higher dimensions. He imagined "bookworms," creatures that lived in only two dimensions, as on a giant sheet of paper. As three-dimensional people, we could peel a bookworm off the sheet of paper, move him to another spot, and lay him back on the paper. To the other bookworms, it would appear that the worm in question vanished and then magically reappeared elsewhere. Or we could slide the bookworm along the sheet, magically "levitating" him. From our perspective of height, we could watch far-distant events in bookworm land, such as what a young worm named Luke is doing. A sealed bottle in bookworm land would simply be the outline of a bottle drawn on the paper. If a secret message had been sealed inside the bottle, we could easily lift it out and put it down elsewhere.

Dr. Michio Kaku tells the story of psychic Henry Slade, who became famous in late nineteenth-century London. Some of the top scientists of the day became convinced that Slade was accessing the "fourth dimension" during his séances to communicate with the dead and perform seemingly impossible tasks. He could take two separate, unbroken wooden rings and intertwine them. He could also remove the contents from a sealed bottle without breaking the seal or the bottle. Such tasks could be accomplished if it was somehow possible to move these objects through a higher dimension, but not in our world. The astounded scientists, of course, had been deceived by a skillful performer. But the fact remains, accessing higher dimensions could allow for seemingly magical powers.

Does this mean Yoda can access higher dimensions to levitate objects and see events at great distances? Dr. Kaku admits that "such feats of 'magic' are, in principle, possible within the realm of hyperspace physics," yet he cautions that "the technology necessary to manipulate space-time far exceeds anything possible on the Earth, at least for hundreds of years."

If the Force somehow provides access to these additional dimensions, Jedi could levitate objects, observe events outside their field of view, and transport objects from one spot to another. They could even drink the beer out of your bottle without opening it.

ZEN AND THE ART OF NERF HERDING

Of all the powers associated with the Force, the one that seems like it would be the most fun is the ability to influence others. "You will take me to Harrison Ford now." If only.

Using that "old Jedi mind trick," Obi-Wan makes a stormtrooper believe R2-D2 and C-3PO are not the renegade droids carrying the Death Star plans, and Luke makes Jabba's chief aid, the tentacle-headed Bib Fortuna, admit him to Jabba's throne room. In both these cases, those influenced repeat the words of the Jedi as if in a trance and then carry out his wishes.

Although this power isn't one we see every day, it does resemble a phenomenon most of us are familiar with, one known since the time of the ancient Babylonians, and studied by scientists for more than two hundred years. Under the right conditions, hypnotists can influence people to say and do things they wouldn't normally say and do.

While in a hypnotic state, a subject is focused yet highly relaxed and suggestible. He becomes detached from his sense of reality, his critical faculties, and the sensory input of his own body. His attention is absorbed and directed by the issues raised by the hypnotist. The hypnotist can then make some suggestion that the subject—unless he finds it extremely objectionable—will believe or follow. Dr. Michael Yapko, clinical psychologist and author of the authoritative hypnosis text *Trancework*, explains, "If someone is ambivalent, riding the fence, you can sway them." If the stormtrooper and Bib aren't passionate about their jobs, then they may be able to be swayed.

Suggestions can make a person behave in a particular way, or see or hear things that don't even exist. A subject may believe a mosquito is buzzing around him. He may stick his hand in a bowl of ice water and find it pleasantly warm, or he may look at a colorful drawing and see it only in shades of gray, as if color-blind. In response to suggestions, he may become paralyzed, amnesic, delusional, or insensitive to pain. Some people have even been able to undergo childbirth or major surgery, such as an amputation, without feeling pain.

Observing such abilities actually inspired Dr. Yapko to study hypnosis. "People have latent strengths and capabilities they aren't aware they have." Some subjects find behavioral changes, such as quitting smoking, weight loss, or overcoming phobias, easier. These latent abilities may also be used for less constructive purposes. Subjects may believe their posteriors have been glued to chairs (I wouldn't mind seeing a platoon of stormtroopers in that condition). Or they may believe R2-D2 and C-3PO are not the droids they're looking for. Dr. Yapko stresses that hypnosis is merely a tool. "Hypnosis is not a good thing. Hypnosis is not a bad thing. It's all in how you apply it." Does that mean hypnosis has a light side and a dark side? "Sure," he answers.

Hypnotists phrase suggestions in certain specific ways. Which type of suggestion is best depends on the individual. Knowing what kind of suggestion to use, Dr. Yapko says, is part of "the artistry of hypnosis." With a direct suggestion, the hypnotist tells the subject what he wants the subject to believe or do, as in "These aren't the droids you're looking for." With an indirect suggestion, the hypnotist presents the desired belief or action in a more oblique way, as in "Isn't it a relief to see that these aren't the droids you're looking for?" Obi-Wan and Luke both give direct suggestions rather than indirect ones. Subjects looking for answers may be happy to adopt a direct suggestion, while those anxious and emotionally guarded may respond better to an indirect suggestion. Luke's suggestion is authoritarian in nature, which as described by Dr. Yapko is, "You will do X," as in "You will take me to Jabba now." Obi-Wan's suggestions are not quite so forceful, phrased in a way almost to make the stormtrooper relax, yet they too are authoritarian, commanding the stormtrooper to adopt his statements as fact. C. Roy Hunter, a hypnotherapy instructor, recommends using present tense for suggestions, as Obi-Wan does. If future tense is used, he stresses the importance of specifying when the action will be done. Luke, in his above command, does just that.

So the suggestions given by Jedi, and the reactions of those under their influence, seem quite similar to those in hypnosis. Yet in *Star Wars*, we don't actually see these people being hypnotized. Could the stormtrooper and Bib be hypnotized so quickly? While most of us think of a hypnotist swinging a watch back and forth, saying, "You are getting sleepy," contemporary hypnotists use a variety of methods to induce a hypnotic state. Most of these work slowly, over perhaps fifteen minutes, gradually deepening the subject's relaxation. Yet a rapid technique is often used by stage hypnotists. This method, described as a shock to the system, is the one most similar to what we see in *Star Wars*. The hypnotist gives a sudden forceful command in a surprising manner, and puts the subject into a hypnotic state in seconds. Roy Hunter explains, "The participant or client will experience a 'moment of passivity' during which he or she will either resist the trance, or 'let go' and drop quickly into hypnosis." Though not loud or showy, both Obi-Wan and Luke make forceful assertions that would certainly surprise those to whom they speak.

Yet Dr. Yapko stresses that stage hypnotists do some preliminary work with subjects before they "instantly" induce hypnosis. "They administer suggestibility tests to deduce a person's level of responsiveness in that situation." One test might be telling a person she has a special glue on her eyelids, and she won't be able to open her eyes. The hypnotist then tells her to open her eyes, and if she can't, he knows he has a good subject for his show. Inducing a traditional hypnotic state in a subject who has never been hypnotized before, even with these rapid methods, takes at least a minute or two.

So it's unlikely that Ben or Luke induce a traditional hypnotic state in their subjects. Yet in many cases, suggestions can be given and accepted without the subject being hypnotized, just like the suggestion of glue on the eyelids. Some therapists call these "non-hypnotic suggestions," while others prefer to define hypnosis much more widely. The difficulty in establishing exactly when someone is hypnotized arises from the fact that we still don't have a clear definition of this state. The EEGs of hypnotized people look much the same as those of people under normal conditions, so we can't definitively say whether someone is in a hypnotic trance or not.

Neuropsychologists disagree about exactly what is occurring in the brain during hypnosis. Scientists offer three general explanations. Dr. Yapko believes that "there are elements of each explanation evident in hypnosis." First, some scientists believe the hypnotic state is a unique, altered state, in which the mind operates in a different way than usual. That would explain why hypnosis allows us to access abilities we can't normally access. Others believe we enter similar states of high concentration every day, when we daydream or become engrossed in a great book or movie, like *The Phantom Menace*. Still others believe that hypnosis is simply a process in which the subject decides to cooperate with the hypnotist, imagining various things and acting out various behaviors. In this view, subjects are suggestible, imaginative, and intimidated by authority figures, whom they don't want to disappoint.

The second and third explanations allow a wider definition of hypnosis, allowing hypnotic activity to occur even when someone hasn't been formally hypnotized. Some therapists consider a conversation including direct suggestion to be hypnosis. Dr. T. X. Barber, author of *Hypnosis: A Scientific Approach*, points out that a conversation involving direct suggestions to experience, think, and feel particular things can stimulate responses quite similar to those obtained in a traditional hypnotic state. Such suggestions might make a subject believe that his extended arm is becoming horribly heavy, or that he can't unclasp his hands. In one study, about 40 percent of participants experienced visual or auditory hallucinations when these were suggested, without any hypnotic induction.

Dr. Yapko believes that "hypnosis occurs to some degree whenever someone turns his or her attention to and focuses on the ideas and feelings triggered by the communications of the guide." This condition can be triggered merely by an offhand remark. If it influences the subject's experience, then it is hypnotic. Rather than entering a unique state, then, the patient is interacting with the hypnotist in "hypnotic patterns of communication" that focus the patient's attention.

For non-hypnotic suggestions to work, Dr. Yapko believes a

hypnotist requires skilled control of his voice and body so that they project credibility and authority. In the authoritarian tech- nique, the hypnotist fixes his eyes on the subject's. Both Ben and Luke do this, drawing their subjects' attention to them. Other methods may also help focus a subject's attention. When Obi-Wan talks to the stormtrooper, he makes a small gesture with his hand. Luke extends a finger toward Bib when he makes his suggestion. Dr. Yapko points out, "When somebody gestures, it draws atten- tion to them and to the gesture. It's a good way of securing the person's attention to a greater extent."

More than that, the hypnotist must seem at all times to know exactly what he's doing. "Certainty is persuasive. When Steven Seagal walks into a room with ten bad guys, and he insults the man in charge, there's something intimidating about someone with that level of confidence. This makes the bad guys afraid. They think, 'Maybe he knows something I don't.' That uncertainty can lead you to be more likely to comply with or be intimidated by that person whose level of certainty is so great." Both Obi-Wan and Luke are quite certain in their assertions.

Yet a hypnotist needs more than confidence and an authorita- tive posture and voice. For rapid induction or non-hypnotic sug- gestions to succeed, according to Dr. Yapko, there must be "some expectation, some history, some rapport, some willingness to ac- cede to an authority." A stage magician may have authority from his fame. The subject believes the magician has some hypnotic abilities. He expects to be hypnotized, and probably wants to be hypnotized. He may have seen demonstrations of hypnosis before, which would provide a history, and his volunteering reflects a will- ingness to submit to the hypnotist's authority. A psychotherapist's credentials and the trust the subject has placed in him give him his authority. The preliminary work between the therapist and the subject sets up a rapport between them.

In the case of Luke and Bib, expectations are set up ahead of time when Bib sees Luke's holographic message, which is pro- jected by Artoo. Luke identifies himself as a Jedi knight, and hints that if Jabba doesn't agree to turn over the frozen Han, an "un- pleasant confrontation" may result. Dr. Yapko, a *Star Wars* fan, says, "If the guard knew that Luke was on a par with Darth Vader in terms of using the Force, and that he was under an implied threat if he countered Luke, that might be enough to intimidate him or make him susceptible to Luke's influence." In this case, some history and some expectation—that Luke might use that "old Jedi mind trick"—has been established.

But in the case of Obi-Wan and the stormtrooper, no such history or expectation could have been established. While the result of Obi-Wan's suggestions is similar to that in hypnosis, it seems he must be taking a shortcut to get this effect. His mind appears able to directly affect the minds of others, as we discussed in connection with quantum mechanics.

We have another example of the Force's influence, though, that does not seem hypnotic at all. A hypnotist works by making verbal suggestions that the subject then accepts. While we might argue that Obi-Wan works this way with the stormtrooper he encounters on Tatooine, he doesn't even speak to the stormtroopers on the Death Star. After deactivating the tractor beam, he needs to make the stormtroopers look the other way to get past them. With a small gesture of his hand, he somehow makes the stormtroopers believe they have heard a sound down the hallway, which makes them look away. No process we can recognize as hypnosis is occurring here. It seems as if Obi-Wan is either telepathically affecting their minds, planting a suggestion there, or else that he has created an actual sound down the hall by the manipulation of matter and energy. As Dr. Yapko says, "It must be the Force."

MEASURING THE FORCE IN THE LAB?

Since the hypnotic state is not easily definable, researchers have focused on other paranormal phenomena to find concrete evidence of psychic powers. Phenomena such as telepathy, clairvoyance, and psychokinesis have been the subjects of extensive parapsychological research.

We discussed earlier whether the Force might be theoretically possible. Yet theory isn't the only issue involved in exploring the Force. Science seldom works by creating a theory and then discovering evidence of it. Usually, science works by observing an unexplained phenomenon and then creating a theory that can describe or explain it. Gravity existed before we understood its workings—even now we don't entirely understand it—and the phenomena associated with the Force may exist, whether or not we can explain how they might theoretically operate. Dr. Jessica Utts, professor of statistics at the University of California at Davis and researcher of paranormal phenomena, says, "There are a lot of things we accept scientifically before we know how they work. Consider learning and memory. Psychologists do not understand these simple human capabilities, yet no one would deny the existence of these phenomena just because we do not understand them."

Since the establishment of the Society for Psychical Research in England in 1882, scientists have been engaged in empirical research to see if they can document paranormal phenomena. Many believe that they have. They have different ideas, though, about how these phenomena relate to traditional science. Some parapsychologists believe that these phenomena may be compatible with our current understanding of the universe. They point to quantum mechanics as a possible source of explanations. Since quantum theory deals in statistical probabilities, and paranormal phenomena—as we will discuss—also appear statistical in nature, they believe the paranormal is caused by quantum effects. Others believe the proof of these phenomena will require a complete revolution in science, in which we throw out everything we know about physics. Dr. Ray Hyman, professor emeritus of psychology at the University of Oregon and self-proclaimed skeptic, believes that "Neither relativity nor quantum theory can cope with a world that harbors the psychic phenomena." Yet others believe these phenomena are truly supernatural and can never be explained by science.

But whether they can ultimately be explained or not, has convincing evidence of paranormal phenomena been found? Many scientists believe their experiments have yielded proof of such powers. Yet the scientific community at large does not believe conclusive evidence exists. Extraordinary claims, they say, require extraordinary evidence, while the evidence of paranormal phenomena remains plagued by flaws and skepticism.

Part of this failure is due to faulty experimental procedures. Experiments designed to detect paranormal phenomena seem particularly prone to cheating subjects or experimenter fraud. Many scientists have been taken in by magicians and charlatans whose techniques of fooling an audience—including hidden radios—are far more sophisticated than the scientists know. Magician James Randi, known as The Amazing Randi, has made it his mission in life to uncover such charlatans and to reveal weaknesses in paranormal research. He does this through the James Randi Educational Foundation of Fort Lauderdale. As someone who has made a living by simulating acts of magic, Randi believes the simulation of magic should never be mistaken for or misrepresented as real magic.

Dr. Hyman relates one instance when he himself was fooled. The government asked him to look into the work of a scientist investigating a woman who could allegedly read books just by running her fingertips over the pages. "She seemed a very ethical person. The woman didn't want any publicity. She didn't even want to be tested. I knew something was wrong, but I didn't believe any fakery was going on. It turned out she was cheating this scientist blind." Randi believes many scientists are too accepting of any evidence that validates their beliefs, which makes them vulnerable to deception by others and even by themselves. "I can go into a lab and fool the rear ends off any group of scientists."

Even if all parties are honest, a desire to believe on the part of the scientists can allow design flaws to creep into experiments. If an experiment is not designed very carefully, scientists may convey subtle, unconscious cues to their subjects that tell them the "correct" answer to give to prove their psychic abilities. Once scientists have collected data, they may manipulate it in different ways to find a statistically significant result. And if they can't find a significant result, parapsychologists tend not to report their experiments, reporting only those that yield positive indications of paranormal phenomena. Since the vast majority of scientists who perform paranormal research are those who believe in paranormal phenomena, this problem is hard to avoid.

In addition to these problems, parapsychology differs from other sciences in a number of ways that make scientists from other disciplines hesitant to accept potentially positive results. Most scientific disciplines have basic experiments that can be easily replicated and produce consistent results. These are the experiments your teachers always made you do in science class. Dr. Hyman cites an example. "Isaac Newton did experiments with prisms to show that light can be separated into colors. We can still do those experiments and get those results. Every field of science has hundreds of these textbook-type experiments, which students can do with predictable results. There isn't one experiment in parapsychology like that at all." Dr. Utts explains why. "I can name other phenomena for which students could not be expected to do a simple experiment and observe a result, such as the connection between smoking and getting lung cancer. What differentiates these phenomena from simple experiments like splitting light with a prism is that the effects are statistical in nature and are not expected to occur every single time. Not everyone who smokes gets lung cancer, but we can predict the proportion who will. Not everyone who attempts telepathy will be successful, but I think we can predict the proportion who will."

Another difference between parapsychology and other fields is that paranormal phenomena can't be disproven. In a field like physics, experiments are set up so that the results either support or disprove a theory. If the results disprove the theory, then the theory is discarded. But if a parapsychological experiment fails to show results, Dr. Hyman says, the experimenters will argue that "the conditions weren't conducive to psychic phenomena." Since the workings of such phenomena are wholly unknown, the factors that may affect them are unclear.

All these issues end up making belief a critical component in how you view the experimental results. When these experiments are reviewed by other scientists, those who already believe in psychic phenomena will find the evidence conclusive, while those who don't believe in psychic phenomena won't. Marc Millis believes the only way to approach the data is with an open mind, but he explains, "Open-mindedness goes two ways. You have to be open-minded to the possibilities that things might work, and open-minded to the possibilities that things might not work. A lot of people are only open-minded one way."

What conclusions are we to draw from such research? Many scientists believe that any experiment showing positive indications of psychic phenomena must be flawed, since such phenomena are impossible. Dr. Puthoff feels "Many skeptics are true believers in the impossibility of it. They reject the results because they just can't imagine how it could work. That level of skepticism is essentially worthless." Others, who have attempted to be open minded as Millis suggests, are discouraged by the lack of convincing evidence.

Yet some parapsychologists believe that experiments have proven that paranormal phenomena exist beyond any doubt. Dr. Utts asserts that the skeptics have been unable to explain away significant results. "And it's not for lack of trying." And occasionally a physicist, or an engineer, will be converted by a compelling piece of evidence or a particular study and begin to research paranormal phenomena. Yet believers find their evidence is not widely accepted. "Parapsychologists are the Rodney Dangerfields of science," Dr. Hyman says. "They want to get respect. Some of the disrespect they get is not earned, and some is."

From their view, parapsychologists are struggling with meager financial resources to capture magic in the lab. The effects of psychic phenomena are so small, Dr. Utts admits, that it takes hundreds or thousands of trials to measure a significant result. Also, if subjects are not completely at ease, parapsychologists believe, their psychic abilities might be driven away. That's why so many psychics, when brought into the lab, fail to produce results. If too many controls are put on them, they'll feel distrusted. And indeed, over the history of parapsychological research, as experimental designs are improved and controls are tightened, any psychic ability measured keeps getting smaller and smaller. While cynics feel this indicates the actual ability is zero, believers feel additional controls kill the subject's spirit. The presence of disbelievers may also inhibit a subject's psychic powers, which is one reason that successful experiments reproduced by skeptics often don't generate positive results. Just as Yoda tells Luke he must believe for the Force to work, parapsychologists feel belief is an important component in their experiments.

Now that you know some of the factors involved in psychic research, let's take a look at some of the results. Then you can decide for yourself whether or not the Force is with us.

REACH OUT AND TOUCH SOMEONE

Near the end of *The Empire Strikes Back*, Luke is hanging from what looks like an antenna on the underside of Cloud City. Using the Force, Luke calls out with his mind to Leia for help. She "hears" his call and brings the *Millennium Falcon* to his rescue. As the *Falcon*, now carrying Luke, attempts to make the jump to hyperspace, Darth Vader calls out mentally to Luke via the Force. Unfortunately, you can't "hang up" on a call carried by the Force.

Here the Force allows telepathic communication between people. Parapsychologists like to call telepathy "anomalous information transfer," since information passes from one person to another without any apparent means. Have they found any evidence that we might be able to mentally communicate with each other?

The most interesting results have come from a series of experiments begun by Dr. Charles Honorton and then continued after his death a few years ago by Dr. Robert Morris at the University of Edinburgh, and by other scientists at a variety of institutions.

Dr. Honorton designed a test that he hoped would detect telepathic communication and that would be free of design flaws or opportunities for cheating or fraud. Most scientists agree that, if people do have telepathic powers, they must be very weak for most people under most circumstances. So Dr. Honorton figured that to foster telepathic communication, he needed to free his subjects from all distractions and put them into the ideal state to "receive" communications. Alleged cases of telepathic communication in the past have often been associated with meditation, dreaming, or altered states, which is consistent with what we see in *Star Wars*. Hanging from the underside of Cloud City, Luke goes into a meditative state to call out to Leia. Some experiments have tested subjects for telepathic abilities after either listening to a relaxation tape or a tension-inducing tape. Those who were relaxed performed better than those who were tense. In another experiment, subjects tested stronger for telepathy after meditating.

To make his subjects as receptive as possible, Dr. Honorton put them into a sensory deprivation chamber. They lay on a reclining chair, with half a Ping-Pong ball strapped over each eye. A red floodlight created a hazy visual field (called a Ganzfeld by German psychologists) and headphones hissed white noise into their ears. In this fog without any sharp sensory boundaries, subjects entered a pleasant, altered state within fifteen minutes.

Then Dr. Honorton chose "senders" that were friends or relatives of the subject—to increase the chance of a psychic connection—and put them in a separate, isolated, soundproof and radioproof room. The sender looked at a video clip randomly selected by a computer and attempted to transmit the image telepathically to the subject. The subject gave a running monologue describing any images or thoughts coming to mind.

After the subject came out of the isolation chamber, Dr. Honorton or another researcher would show him four video clips, and ask him to choose which one most closely matched his thoughts while in the isolation chamber. The experimenters did not know which image was shown to the sender, so they could not bias the results in any way.

Since the subjects were shown only four clips, they should have picked the image shown to the sender simply by chance 25 percent of the time, without any telepathy operating. Yet Dr. Honorton's tests showed that the subject picked the correct image 35 percent of the time, much more than could be explained by chance. And the most recent tests run at the University of Edinburgh found the subjects picked the correct image nearly 50 percent of the time. Are these subjects actually receiving telepathic images?

While skeptical scientists have yet to definitively prove the results invalid, they have identified several possible design flaws in the experiment that might lead to misleading results. Remember as you read these that this experiment, like all paranormal research, is set up to measure a statistical effect, not an absolute one. Gravity will make a book fall every time I drop it, but telepathy will not allow communication between two people every time they try it—at least not on this planet. The experiment shows that subjects pick the correct clip more often than would be expected by chance. So any factors that might affect the chances of a subject picking a certain clip are relevant.

One possibility, unintentional correlations may be playing a role in the results. For example, people hearing white noise tend to think about water more often than sex (set your radio to the static in between two stations and test it!). If the video images selected by the computer happen to be of a tidal wave more often than of two people in bed, and subjects more often identify the image of the wave rather than the image of the people in bed, this doesn't mean any telepathic communication has occurred. This simply indicates that a subject listening to white noise is often reminded of water. This was the case in one such experiment.

Another possibility, of four images shown to them, subjects tend to choose either the first or last. If the images selected by the computer happen to be shown to the subject first or last more often than in the middle two positions, the subject might choose them with no telepathy involved. This was a factor in a number of experiments.

If a video clip is played over and over as the sender attempts to transmit the image, the clip might become staticky and degraded. When the clips are played for the subject, then, he could guess which clip had been used.

Another potential danger, researchers may have known which image the computer selected, even though they weren't supposed to. The researchers may have purposely gotten this information, by looking at the counter on the VCR holding the video clips, or they may have subconsciously known it, by the amount of time it took the VCR to rewind after showing the clip.

The researcher might then subtly influence the subject to choose a particular image over the others. In fact, the researcher often spoke to the subject during the selection process, pointing out different elements in the clips and relating them to what the subject said while in isolation. Dr. Hyman questions, "Are you measuring the ability of the subject to pick out the right target, or the ability of the researcher?" In the experiments yielding the 50 percent success rate, most of the successful hits came from trials involving one of three researchers, suggesting the researcher was influencing the results.

So the Ganzfeld evidence remains controversial. You might think that these flaws seem rather small and unlikely. The key issue is whether the flaws identified could account for the results observed. In his study of Ganzfeld results, Dr. Hyman says, "At first I was pretty impressed. But as I kept going through them, I found there wasn't a single experiment in the database of forty-two experiments that was free from obvious flaws." Dr. Utts believes, though, that "if all the experiments have different weaknesses but all come up with the same magnitude of fact, that's pretty convincing." As a physicist and a perfectionist, though, I can't help feeling uncomfortable accepting flaws in experiments.

Whether these flaws can completely account for the effect observed, though, remains unclear.

What tends to convince most believers is not the statistics, but isolated, striking cases that appear to be instances of telepathic communication. In one experimental trial, for example, the video clip used was a scene from the movie *Altered States*. The scene, in shades of red, depicted a hellish hallucination comprised of a jumble of images: a man screaming, people in fire and smoke, a sun with a corona around it, a mass crucifixion, people jumping off a precipice, lava and smoke, and a lizard opening its mouth. Here's what the subject in the Ganzfeld said: "I just see a big X. A big X. I see a tunnel in front of me. It's like a tunnel of smog or a tunnel of smoke. I'm going down it. I'm going down it at a pretty fast speed. . . . I still see the color red, red, red, red, red, red, red. . . . Ah, suddenly the sun. . . the kind of cartoon sun you see when you can see each pointy spike around the sphere. . . . I stepped on a piece of glass and there's a bit of blood coming out of my foot. . . . A lizard with a big, big, big head."

The correspondences are quite compelling. Yet this is only one trial out of many. And the complex video clip offers such a hodge-podge of different images that it seems a subject might stumble into a couple of them by accident. From a scientific perspective, the relevant evidence is not what the subject said, but whether the subject chose the correct clip at the end (which she did). To be convinced by isolated trials, though, scientists must ignore many, many other trials that show no correspondences. While most scientists believe these isolated trials are mere flukes, parapsychologists assert they are examples of a gifted subject caught at a favorable time.

I had my own close encounter with telepathy while writing this book. At the end of my interview with The Amazing Randi, I asked the magician if he could read my mind. He astutely said, "You're thinking it's about time to go and I've been very patient with you."

Who is the most likely to have telepathic abilities? According to parapsychologists, those who score highest believe in telepathy and have training in meditation, relaxation, or creativity. A small group of students at the Juilliard School for the performing arts had the highest rate of matches of a single group, 50 percent. Since Yoda's training incorporates belief, meditation, and relaxation, he may be increasing the telepathic ability of his students. He may want to throw in some creativity training—perhaps a finger-painting class?

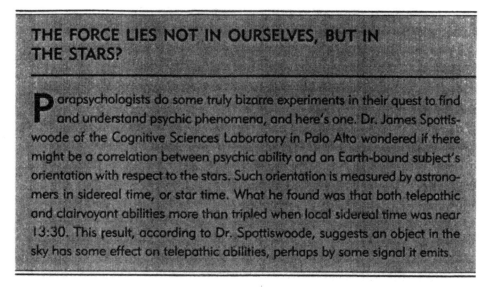

THE FORCE LIES NOT IN OURSELVES, BUT IN THE STARS?

Parapsychologists do some truly bizarre experiments in their quest to find and understand psychic phenomena, and here's one. Dr. James Spottiswoode of the Cognitive Sciences Laboratory in Palo Alto wondered if there might be a correlation between psychic ability and an Earth-bound subject's orientation with respect to the stars. Such orientation is measured by astronomers in sidereal time, or star time. What he found was that both telepathic and clairvoyant abilities more than tripled when local sidereal time was near 13:30. This result, according to Dr. Spottiswoode, suggests an object in the sky has some effect on telepathic abilities, perhaps by some signal it emits.

PEEPING JEDI

Yoda tells Luke that through the Force he can "see" events occurring at a great distance, or events in the past or future. Yoda himself admits to "watching" Luke for a long time. Might we call him a peeping Jedi?

Traditionally this ability to see distant locations and actions has been known as clairvoyance, yet scientists again have come up with their own name: remote viewing. The Pentagon's Defense Intelligence Agency shocked scientists when it revealed, three years ago, that it had been conducting research on remote viewing for the past twenty years. The results, as we might expect, were controversial.

The remote-viewing experiments were set up in several different ways. In one, five sites in the surrounding San Francisco Bay area were selected. A researcher chose one of these sites and traveled to it, serving as a "beacon" to draw the subject's attention to the spot. In a sealed room, the subject sat with another researcher, waiting for word that the "beacon" had arrived at the site. After receiving word, the subject attempted to describe and draw the site. His comments were tape-recorded and later transcribed. The subjects were discouraged from naming objects, such as saying "I see Jabba's palace." Instead they were encouraged to simply describe or draw shapes. This technique brought more successful results, which researchers explained with the belief that psychic talents do not involve the verbal part of the brain. Dr. Utts explains, "What seems to come through are more general images and shapes than high-level interpretation."

In other trials, the subjects were given coordinates of longitude and latitude and asked to draw what was there, without a "beacon." The results were the same with or without a beacon, which brings into question how remote viewing works. With a beacon, remote viewing might not be much different than telepathy, the beacon sending information to the subject. But with no beacon, the subject seems to be somehow traveling mentally to the site and observing it. The remote viewer seems equally successful at short and long distances, and even sometimes sees into the past or future. Some researchers, including Dr. Utts, actually believe remote viewing works through precognition. Since the subject is taken to the actual site after the trial, the subject's earlier description may have been obtained by looking into the future to see it.

Once all five sites had been "remote viewed," the number of successful matches had to be evaluated. The director of the study tried to match the subject's drawings and descriptions to five different sites. Simply by chance, the director would be expected to make correct matches 20 percent of the time. Similar to the Ganzfeld experiments, the results were higher than could be explained by chance.

After twenty years of study, the CIA hired two researchers to evaluate the evidence collected. One, Dr. Utts, found convincing and conclusive evidence of remote-viewing ability in some of the subjects. The other, Dr. Hyman, did not find the evidence convincing. Dr. Puthoff, who was the founder of the research program, says that Dr. Hyman "admitted that the data were statistically significant. But he said there must be flaws, even though he wasn't able to find what they were." Dr. Hyman, though, explains that while the experiments seemed better designed than previous ones, they introduced some new procedures that hadn't been tried before. "Every time new procedures are introduced, it takes time to find the bugs in them. There are always flaws in every experiment." The CIA concluded that the study provided insufficient evidence and no scientific grounds for its claims, and stopped funding the research.

Perhaps we can identify some of the possible flaws in the remote-viewing research by looking at a similar experiment. Drs. David Marks and Richard Kammann at the University of Otago in New Zealand, authors of *The Psychology of the Psychic*, tell the fascinating story of their attempt to reproduce the remote-viewing results of the government researchers. They followed the same procedures and were quite surprised and disappointed to find that they obtained no significant results. All of their subjects failed miserably as remote viewers.

The odd thing was, almost invariably, when the subject and the beacon visited the site after the trial, they would be convinced that the subject had described the site accurately. They would see strong correspondences between the subject's description and various elements of the site, just as we saw correspondences between the scene from *Altered States* and the Ganzfeld subject's description. But when a judge completely unconnected with the experiments would try to match the transcripts and drawings with the list of sites, the judge would draw different comparisons and correlations. He, too, would be convinced that the subject had successfully viewed the sites, but he would connect the wrong transcript to the wrong site.

For example, one site might be a house. The subject and researcher might visit the house and find that the view looking away from the house toward the street matches some of the elements in the transcript corresponding to this site. The judge might instead find the view looking toward the house matches some of the elements of another transcript. Drs. Marks and Kammann call this phenomenon "subjective validation," which they say occurs "when two unrelated events are perceived to be related because a belief, expectancy, or hypothesis demands or requires a relationship." If someone believes a transcript accurately describes a site, and he "visits a location with that description fresh in his mind—any description—he will easily and effortlessly find that the description will match." In other words, "Any description can be made to match any target." Since subjects were discouraged from specifically naming objects, this only made the transcripts easier to interpret in different ways.

If that is so, though, how did the original researchers obtain better-than-chance levels of matching? What allowed their judge to match the correct transcript to the correct site so many times? Well, Drs. Marks and Kammann found that the judge in the government experiments was given a list of the sites in the exact order they were used. This wouldn't matter, unless the judge also knew the order in which the transcripts and drawings were generated. If he did, then simply matching the first site with the first transcript, and so on, would be easy enough. But did he know?

In examining five transcripts of subject Pat Price, a star in the government remote-viewing experiments, Drs. Marks and Kammann found that each transcript contained cues about the order in which they had been recorded. For example, in one transcript Price expresses anxiety about his ability to remote view, from which one might conclude that this is his first attempt. In his second viewing, he mentions this is the "second place of the day," and in his third he mentions "yesterday's two targets." In his fourth, the experimenter says, "Nothing like having three successes behind you." In his fifth, Price refers to the marina that was site number 4. If the judge knew in which order the sites had been visited, he could easily match the transcripts to the sites with 100 percent accuracy, which Marks and Kammann did. The specific description given in the transcript and whether it matched the site would be completely irrelevant.

In other less dramatic cases, the judge could at least re-create a partial chronological order, increasing the chance of a correct match. Some transcripts even had dates on them! Marks and Kammann were unsuccessful in generating a high number of matches in their experiment because they edited such cues out of the transcripts before the judges saw them.

Dr. Puthoff disputes the claim that matches arose because of cues in the transcripts. "To address these criticisms, we turned over the entire experiment to an independent third party. He edited out all the cues and arranged to have the transcripts judged again. The judge made the same matches as originally found in our study, proving that the Marks and Kammann cue hypothesis was false." Dr. Puthoff stresses that accurate descriptions of the targets generated the matches, not the cues.

In examining the transcripts of the government tests, Marks and Kammann found another possible cause for the positive results. They claim that some experimental trials were conducted and not reported, suggesting the government experimenters selected the most promising results to report and discarded others.

The work of Marks and Kammann may help to explain how higher-than-chance matching occurred. But researchers again point to isolated cases that appear to offer compelling evidence of remote viewing. One subject, the same one whose transcripts were studied for cues, was particularly successful. This subject, Pat Price, a former police commissioner, claimed that he would listen to calls on the police radio reporting crimes, psychically scan the city, and then send a police unit to the place where he saw a frightened man hiding. In one remote-viewing trial, Price described a site, a swimming pool complex, in great detail. All the details were accurate except for one: he described water storage tanks at the complex and there were none. Much later, researchers discovered that such tanks had existed in that location in 1913, and so they cite this as a powerful example of remote viewing of the past. While this may be so, it seems to me that if you consider what has been at any site through the entire course of history, you could probably find something to match Price's drawing.

In another experiment, Price was asked to remote view a Russian research and development facility ten thousand miles away. Price drew a crane and a section of a giant metal sphere. Only several years later did the experimenters learn that these actually existed at the site. Yet Price drew many other things that he "viewed" at the site, and these were not found there.

Another remote viewer, Ingo Swann—called Obi Swann by some of the researchers—decided to remote view Jupiter before NASA's Pioneer 10 flyby. Swann, a career psychic, "saw" a ring around Jupiter before the ring was discovered.

In addition to the experiments we've discussed, the Pentagon has regularly consulted remote-viewing "spies" to gather intelligence. Such remote viewers have drawn pictures of secret Soviet submarine construction sites, North Korean tunnels, and the

house where American general James Dozier was held while kid-
napped in Italy. Their descriptions and drawings, when compared
with reconnaissance photos or other data, are claimed to be accurate 15 to 20 percent of the time.

How could such success rates be obtained, if remote viewing
is not possible? Magician James Randi provided a fascinating example as we spoke on the phone one day. I was in New Hampshire,
Randi in Florida. Randi said he would "remote view" something
that I drew. He asked me to draw two simple geometric shapes,
one inside the other. I grabbed a pen and quickly drew the two
shapes. (If you'd like to play along, stop reading now and draw
your own two shapes. Ready?) He then asked if I'd drawn a trian-
gle within a circle. I was amazed. That was exactly what I'd drawn!
(How about you?) I asked how he had guessed that. He answered,
"This is one of the tricks of the trade. We know what people are
likely to draw under certain circumstances." Most people, when
asked to draw one geometrical shape inside another, draw what I
drew.

That explains some of the positive results, but how might a
drawing by a "remote viewer" agree with reconnaissance photographs of an area the viewer has never seen? The key, Randi tells
me, lies in drawing a very general shape that can be interpreted in
many different ways. If the researcher wants to see a correlation
between the drawing and the reconnaissance photograph, then he
will find one, just as Marks and Kammann described with "subjective validation." As an example, Randi says to draw a narrow, horizontal ellipse, like a worm crawling across your paper. Then from
each of the two narrow ends of the ellipse, draw a slanting line
upward so that the two lines intersect and form an inverted V
above the ellipse. This one drawing, Randi points out, could be
interpreted in a number of ways: as an ice cream cone, an old-
fashioned cup from a vending machine, a searchlight, a tent, a
sailboat, a horn, a party hat, and many more. As Randi says, "It's
close enough for government work."

Do these explanations account for all the positive results
found? As with much paranormal research, it's hard to say. And
in the case of remote viewing, much of the research remains classi-
fied. Both Dr. Utts and Dr. Hyman agree that more investigation

is warranted, though the government seems unwilling to continue funding the research.

Yet Randi's foundation offers a reward of $1.4 million to any remote viewer who can tell Randi what he has sealed within a specific cupboard in his Fort Lauderdale home. The cupboard holds a simple household object, such as an orange or a bottle of ketchup—changed every few weeks for security purposes. No one has met this challenge. Those who claim to be remote viewers, says Randi, "make all sorts of excuses. They would rather view Lebanon or Jupiter than Fort Lauderdale, because they can fool their clients into believing that they are correct, but they can't fool me."

This reward is not only available to remote viewers, but to anyone who can exhibit a true paranormal ability under controlled conditions. If you're feeling the Force and want to try for the reward, you can get more details on Randi's website, which is listed under "Recommended Reading" at the end of this book.

THE REVENGE OF THE SCIENTISTS

In the introduction to this book, I said that my purpose was not to nitpick. But I thought I'd give a few scientists this one chance to share the *Star Wars* science goofs that most got their goats.

Marc Millis, leader of the breakthrough propulsion physics program at NASA: "If you can pick up an object and hurl it at Luke Skywalker, why not just hurl Luke Skywalker? Why not bend the light saber out of the way as it's coming at you? Obviously there's a limit as to how well this thing works."

Dr. Victor Stenger, professor of physics at the University of Hawaii: "This business where they travel great distances instantaneously. I don't buy it. I think wormholes exist, but I don't think people will ever be able to go down there and end up in another part of the universe in any controlled way. Maybe particles can do that, but a whole composite object would be torn apart."

Dr. Miguel Alcubierre, a researcher at the Max Planck Institute for Gravitational Physics in Pottsdam, Germany:

"The sound of explosions in space. That bothers me most. Something blows up, and you hear a big ka-boom! Explosions would never make a sound. Being a science fiction fan on the one hand and a physicist on the other, there's a duality. You know as a physicist that most of the things that you like as a fan can't be done. You have to turn off your brain. Sometimes you're just annoyed that something you know we can do, they do completely wrong."

Dr. Clifford Pickover, biochemist and author of *The Science of Aliens*: "The various alien musicians in the Max Rebo Band seem to appreciate music exactly in our audio spectrum. It seems unlikely that music would evolve so similarly."

Mark Stafford, engineer, State of Alaska Division of Emergency Services: "The fighters in the movies bank when they turn, but airplanes do that because the airfoils that give them lift make them tilt when they turn. Ships in space should just turn on an even line. And why didn't Grand Moff Tarkin just blow Yavin up at the end of *A New Hope*, instead of taking the time to go around it? Blowing up Yavin would almost certainly destroy all its moons."

Dr. Jessica Utts, professor of statistics at the University of California at Davis: "This isn't a goof, but Obi-Wan Kenobi's name has a play on words. Obi is pronounced the same as OBE, out-of-body experience. It's a little message that it is possible that one can have an OBE."

Steve Grand, director of the Cyberlife Institute: "I try never to let scientific implausibility get in the way of a good story! Ordinary terrestrial movie plots are implausible enough, so why complain about goofs in Sci-Fi?"

A JEDI PICK-ME-UP

The Force allows a Jedi to move objects just by thinking about it. Darth Vader uses this ability in his light saber duel with Luke, mentally tossing pieces of equipment at him; Yoda even raises Luke's X-wing fighter out of the swamps of Dagobah.

People have claimed to perform similar acts here on Earth. The ability to move objects without any known physical means is called psychokinesis. As with both telepathy and remote viewing, psychokinetic power is believed to be rare and small. In fact, most parapsychologists feel psychokinesis is probably limited to the quantum level. So lifting an X-wing seems unlikely at this point.

The most promising results come from Dr. Robert Jahn's experiments, conducted at the Princeton Engineering Anomalies Research Laboratory. Dr. Jahn, the former dean of the School of Engineering and Applied Science at Princeton University, became interested in psychokinesis when a senior student's project seemed to find evidence of it on the quantum level. Intrigued by the possibility that consciousness might be "creating reality" on this level, Dr. Jahn followed up with experiments of his own.

Many of these experiments involve a random number generator (RNG), which we discussed earlier. In this case, the possible numbers are limited to ones and zeroes, so that the RNG performs the electronic equivalent of tossing a coin, with a 50 percent chance of a "heads" outcome and a 50 percent chance of "tails." The generator can perform 200 coin tosses per second, a speed that allows Dr. Jahn to collect massive amounts of experimental data.

Dr. Jahn has a participant sit in front of the generator and try to influence the outcome, trying to make the RNG produce either more heads or more tails. If the subject has no effect, then the sequence of heads and tails should be random and each should appear 50 percent of the time.

When Dr. Jahn averages the results of many trials, he finds a very small effect, yet one he claims is statistically significant. In one typical series of experiments, when participants tried to generate more heads in the 200 tosses, the average number of heads was 100.037, while the number of tails was 99.966. Thus when subjects want a flip to come up heads, the chances of it turning up heads are not 50 percent, as we would believe, but 50.02 percent. In every 1,250 flips, heads will come up one extra time. This may sound like a tiny, tiny effect, and it is.

Certainly if you flip a coin two times, you won't always get one head and one tail. You could easily get 100 percent heads and 0 percent tails, and we wouldn't need psychokinesis to explain that. If you flip a coin 20 times, you won't always get 10 heads and 10 tails. Yet with 20 flips, we would consider it quite odd if you flipped 100 percent heads. With this larger number of trials, such a huge deviation from chance is very unlikely. If you flip a coin 200 times, you won't always get 100 heads and 100 tails. Yet we'd now consider it completely spooky if you flipped 100 percent heads. We'd have to check you for demonic possession—but that's another book. The result of all this is that the larger the number of flips, the smaller any deviation from chance should be. Since any deviations should average themselves out over the long haul, large numbers of flips should generate results very very close to 50–50.

Over the past twelve years, Dr. Jahn has tested 91 different subjects in 1.7 million trials, for a total of 340 million coin tosses. At that level, having subjects flip 50.02 percent heads is a significant deviation from chance. The possibility of such an outcome occurring by chance is one in 2,500. Dr. Jahn believes this proves that people have the ability to affect sensitive electronic components with their minds.

To see if people could affect larger-scale objects, he built a large "random mechanical cascade," which looks like a cross between a carnival game and a pinball machine. Nine thousand polystyrene balls, about half the size of Ping-Pong balls, are dropped down through the ten-foot-high six-foot-wide cascade, which has 330 pegs that send the balls bouncing in random directions. At the bottom of the cascade is a row of nineteen bins. Usually most balls land in the center bins and fewer in the outer bins, the heights of the stacks of balls forming a bell-shaped curve. Dr. Jahn has participants try to influence the paths of the balls, making more go to the bins either on the left or the right. He has found that participants are able to make more balls land to the left, though not to the right, for some reason he has yet to understand. The possibility of the measured results occurring by chance is just one in 33,000.

Just like the other paranormal phenomena, psychokinesis seems to violate everything we know about physical forces. Dr. Jahn has performed experiments where participants try to influence the results from great distances, as far as Russia and New Zealand. Those participants have as strong an effect as those in the lab. And when participants try to influence the results days before or after the test is run, they too have as strong an effect, violating the principle of causality.

Dr. Visser finds the larger-scale psychokinesis particularly difficult to accept. As we discussed in connection with propulsion, one of Newton's laws of motion is that every action has an equal and opposite reaction. Dr. Visser points out, "If you push something, it will push back. I find it difficult to understand how you could hope to do something like telekinesis on heavy objects without causing some back reaction that could damage that gray matter in your skull."

Dr. Jahn's results remain controversial. Some scientists believe that scientific bias has played a role in these experiments, and that Dr. Jahn has manipulated the data to yield the results he wants. Others feel that the effect measured is so small, there are many possible ways of explaining it. Still others believe his experiments are flawed in some way. For example, participants are left alone in the room with the random number generator and might do something to affect it, either intentionally or unintentionally, such as kicking the machine, waving a magnet near it, or even leaning toward it and creating some small static electrical effect. Dr. Jahn replies, "It's silly to think that a lab working on this for twenty years, that has the list of achievements we do, would leave itself open to such nonsense. Our experiments include many layers of protection."

Yet a 1988 government review of Dr. Jahn's work found that just one participant, responsible for 15 percent of the trials, is responsible for *half* of the successful outcomes measured. Dr. Hyman believes that this participant is the laboratory manager and the actual designer of the computer program. "For me it's problematical if the one who runs the lab is the only one producing the results." This may explain why no other lab has been able to reproduce Dr. Jahn's results. The government review concluded that if the subject's trials are removed from the data, the experiments yield only a tiny effect, one more possible to have occurred by chance. Yet Dr. Jahn counters, "The report is desperately outdated. It's also simply not true. If you look at the figures, there are no spikes or superstars."

Dr. Utts, who believes remote viewing is possible, is skeptical about the ability of the human mind to move objects. She has an alternate theory, though, to explain the phenomena. In Dr. Jahn's experiments, participants decide when to press the button to begin the trial. Dr. Utts believes that participants can look into the future and know when the RNG is going to happen to put out a greater than average number of heads. As we know, over a small number of flips, significant deviations from chance can occur. When asked to produce more heads, then, the participant waits until that moment and presses the button. In this way, she believes, remote viewing and psychokinesis work in the same way, through precognition.

Even if such abilities do exist, they aren't on the level of those associated with the Force. Unfortunately, you won't soon be throwing pieces of equipment at your enemies or lifting droids with your mind. Dr. Jahn, though, believes we might someday build machines sensitive to our minuscule mental efforts, machines that might respond by carrying out some task for us. In the meantime, you still might be able to influence when radioactive decay occurs. And this may come in handy if you find yourself in a sealed cell with a blaster aimed at you, and that blaster will be triggered by a radium atom's radioactive decay. . . .

IGUANA AND JEDI MASTER?

A few of my favorite parapsychological experiments explore the following compelling questions:

- Why do slot machines tend to give huge payouts during full moons?
- Can humans affect the growth of algae with their minds?
- Do cockroaches have psychokinetic powers?

Since Star Wars aliens are able to access the Force, we might expect that animals can as well. One bizarre study tested baby chicks for psychokinetic abilities. The experiment used a self-propelled robot built by Dr. R. Tanguy, which looked rather like a small Artoo unit. It had two wheels and one fixed leg, and was driven by a random number generator. The random numbers told the robot how to move its two wheels. The wheels, controlled by separate motors, might move in the same direction, driving the robot forward or backward, or they might move in opposite directions, rotating the robot clockwise or counterclockwise. The robot was constrained to stay on a rectangular surface three feet by five feet and move randomly around on it.

Dr. Rene Peoc'h, working with the French Foundation Marcel et Monique Odier de Psycho-physique, wanted to test whether the chicks could influence the movement of the robot. Chicks like light during daytime hours, and they will cry out if suddenly put into darkness. Dr. Peoc'h put the chicks in a box on the rectangular surface with the robot. The chicks were put on the right side of the surface. A lighted candle was put on top of the robot and the lights turned out. Could the chicks influence the robot to bring the candle near, banishing the darkness?

Dr. Peoc'h measured how much time the robot spent on the left half of the board versus how much time it spent on the right. In 71 percent of the trials involving the chicks, the robot spent more time on the right half of the board, nearer the chicks, than the left half. In trials with an empty box and no chicks, the robot spent equal amounts of time on either side of the board.

After reading this, I toyed briefly with the idea of testing my iguana, Igmoe, for telekinetic powers. I came up with the perfect experiment: mount a heat lamp on top of a robot and see if the heat-loving Igmoe can make the robot bring the lamp close for his comfort. Then I realized Igmoe already has complete control over his heat comfort. I am the robot, responding to Igmoe's every desire, adding heat lamps, taking away heat lamps, letting him outside and then back inside. Perhaps he does have the Force. But does that mean I'm weak-minded?

I mentioned earlier a 1988 government review of Dr. Jahn's work. This review actually studied a much wider range of research, encompassing all the best evidence for paranormal abilities. The survey was conducted by the National Research Council at the request of the U.S. Army. If any paranormal abilities exist, the Army wants their soldiers to have them. After all, soldiers are told to "Be all that you can be." In discussing the possible applications of paranormal powers, the report mentions "intelligence gathering, . . . planting thoughts in individuals without their knowledge." Authors of the report also envision a " 'First Earth Battalion,' made up of 'warrior monks,' who will have mastered almost all the techniques under consideration by the committee, including the use of ESP, leaving their bodies at will, levitating, psychic healing, and walking through walls." Sounds like the Jedi Knights, doesn't it?

Unfortunately, the report concluded that there was "no scientific justification from research conducted over a period of 130 years for the existence of parapsychological phenomena." Dr. Hyman, who participated in the review, agrees. "There's a lot of evidence. It may not be good evidence, but it's evidence."

Yet Dr. Utts believes, "We have proven it. But I don't think we'll get out of this impasse with skeptics until we come up with an explanation. And I understand that. As a statistician, I look at data differently than a physicist. I think we've got the data, we've got the proof."

Even skeptics like Dr. Hyman, though, "leave the door open. This issue of whether psi is real or not is not going to be settled in my lifetime." And the National Research Council, despite the negative findings in their report, recommend that future research in psychokinesis, remote viewing, and telepathy be monitored by the Army. Just in case.

An experiment that proves once and for all whether psychic powers exist remains elusive. Yet it may be that someday these controversial experimental results will be taken for granted, perhaps explained, and maybe even controlled. Yoda would prefer it that way.

While Earth has not yet provided evidence of a power like the Force, we may be measuring phenomena that defy our current understanding of the universe. Even if such results are explainable by experimental error or other factors, though, the truth remains that at the quantum level, while we can describe what occurs, we have no idea why it occurs or even how it occurs. These mysteries remain to be understood, and in finally understanding, we will discover fascinating and bizarre new truths about the world in which we live, as well as yet more mysteries. That is the excitement and the challenge of science.

While science remains a long way behind "a galaxy far, far away," we've made great strides in catching up with George Lucas's vision in the twenty-two years since *Star Wars* first arrived on movie screens. A vision of a universe filled with planets and aliens that seemed impossible then seems quite possible now. And technology that seemed outlandish and physically impossible now seems reasonable and theoretically within our grasp. Perhaps in a few more thousand years, we'll be living in a world much more like *Star Wars* than we ever dreamed.

AFTERWORD

There was no father. I carried him. I gave birth.
I raised him. I can't explain what happened.

—Shmi Skywalker, *The Phantom Meance*

I mentioned in the introduction that this book was written before *The Phantom Menace* was released in movie theaters. I was
able to include discussions of many elements from the movie,
including the planet Naboo, Jar Jar Binks, the double-bladed light
saber, and the STAPs, because of advance information I gathered
about *The Phantom Menace*. Even with that advance information,
though, several elements in the movie took me by surprise.

As the paperback edition goes to press, I have the unique opportunity to add some further commentary. I've seen *The Phantom
Menace* a number of times now—more than I want to admit in
print—and it introduced several striking new pieces to the *Star
Wars* universe. By far, though, the most surprising, fascinating,
and controversial new element is the midi-chlorians. Wherever I
go, *Star Wars* fans ask me how realistic the midi-chlorians are, and
whether they might make the Force more scientifically plausible.
The problems and possibilities inherent in the Force, which we
discussed in Chapter 5, still stand. But the midi-chlorians add a
new wrinkle to the Force, presenting their own problems—and
possibilities.

GOOD THINGS COME IN SMALL PACKAGES

In the movie, we learn that the midi-chlorians are microscopic life forms that reside within the cells of all living things and communicate with the Force. In *The Phantom Menace* novel, we learn more. The midi-chlorians have a collective consciousness and intelligence. Without the midi-chlorians, life could not exist. They provide something critical to all life forms, and in turn, all life forms provide them with something they need as well. Thus the midi-chlorians are said to exist in a symbiotic relationship with their hosts, each benefiting from the other.

While we have nothing exactly like the midi-chlorians in our universe—if only we did!—perhaps the closest we can come are mitochondria. Mitochondria are sausage-shaped microscopic structures within our cells. These structures, or organelles, exist in every cell in your body, and in every cell in every animal, plant, and fungus—all macroscopic life on Earth. They are found in the cytoplasm outside the cell nucleus, and they carry their own DNA, their own genes, separate from the DNA in the cell nucleus.

You probably remember your biology teacher calling mitochondria "the power plants of cells." The mitochondria produce the biochemical energy that the cell uses to carry on its life processes. In a process called respiration, the mitochondria take in sugars, fatty acids, and amino acids from the food you've eaten, and use oxygen to break these ingredients down and release energy. The mitochondria then produce adenosine triphosphate (ATP), trapping this energy in a form where it can easily be released and used. ATP is the fuel the cell uses to make proteins, fats, carbohydrates, and nucleic acids. It helps the cell transport substances into other cells; it makes your muscles contract every time you move; and it even helps send nerve impulses from one cell to the next. Mitochondria develop in many different shapes and sizes and quantities, depending on the type of cell they are in. A cell that requires a lot of energy, such as a muscle cell, will have many mitochondria. An average cell contains hundreds of them.

Thus mitochondria exist within our cells, as do midi-chlorians; they exist in many different life forms; and they provide something important to us—energy. How did mitochondria come to develop and be so widespread? If we can answer that question, then perhaps we can answer the next. How might midi-chlorians have developed and become a part of every living thing?

Just as scientists investigate how man evolved, they also investigate how our cells evolved into the complex structures that they are, and how mitochondria came to be part of them. Most scientists now believe that mitochondria developed from a precursor that lived independently. Because mitochondria have many characteristics in common with bacteria, scientists believe this precursor or ancestor was a bacterium. This bacterium was very talented. It had a very efficient system for producing energy and storing it as fuel. According to one theory, about 1.5 billion years ago, a primitive amoebalike single-celled organism engulfed and incorporated this bacterium into itself. These two life forms grew to live in a symbiotic relationship, each gaining advantage from the other. The cell profited from the bacterium's ability to trap energy released during respiration and create fuel. The bacterium benefited from the stable environment and nutrients supplied by the host cell.

As their relationship flourished, the host cell gradually took over the replication and maintenance of the bacterium, allowing the two partners to better coordinate their efforts. Many genes within the bacterium were lost because they were no longer necessary; others were transferred to the nucleus of the cell. Although the bacterium retained some of its genes, which helped it to continue producing fuel, the bacterium gradually lost its ability to function independently and evolved into the mitochondrion we know today. This loss of independence is not unusual, since most organisms in symbiotic relationships cannot exist without each other.

We can imagine midi-chlorians developing and becoming part of the cells of many organisms in much the same way. Just as the bacterial ancestor of mitochondria formed a symbiotic relationship with a primitive cell, so could the ancestor of the midi-chlorians. It could provide something beneficial to the cell, while the cell provided a beneficial environment for it.

How likely is such a development? If the bacterial ancestor of the midi-chlorians did develop on a planet, would that bacterium be engulfed by a primitive one-celled organism and gradually evolve into midi-chlorians? In Chapter 2 we discussed various evolutionary developments, and estimated their likelihood by looking at the number of times they arose independently on Earth. So how many times did mitochondria arise independently on Earth?

While scientists are still not sure exactly which bacterium was the ancestor of mitochondria, most are currently united in the belief that mitochondria arose only once. All mitochondria in all complex organisms developed from one incident, in which one primitive cell engulfed one bacterium. That suggests this event is not likely. Dr. Gray, professor of biochemistry and molecular and evolutionary biology at Dalhousie University, finds it very unlikely that life on other planets would contain mitochondria, yet he does believe alien life will have "some kind of host/symbiont relationship. That seems to be a hallmark of life on Earth. There are very few organisms—I can't think of any in fact—that don't associate to some degree with other organisms and aren't dependent on other organisms."

The midi-chlorians, however, are said to exist in the cells of *all* living things. Qui-Gon says, "Without the midi-chlorians, life could not exist." That means not even a primitive one-celled organism could exist to swallow up the ancestor of the midi-chlorian, as we've postulated above. If we accept that no life can exist without midi-chlorians, though, we have a big problem. How did the midi-chlorians develop in the first place? Unless they suddenly appeared full-blown, they would have had to evolve gradually from something more primitive. Yet that something more primitive could not survive without midi-chlorians.

So I'd like to suggest a slight alteration to Qui-Gon's statement. Perhaps he meant that without the midi-chlorians, multicellular life could not exist. As we know, when you're talking to a Jedi, the truth can be a little slippery. If this is what Qui-Gon really meant, then mitochondria are again a perfect comparison. Mitochondria are not in the cells of *all* living things, since bacteria produce their required energy without them. But all complex life forms on Earth have mitochondria. Dr. Michael Gray says, "The vast majority of higher life depends on mitochondria. Multicellular beings could not get by without them." And so it seems reasonable that we could claim the same for midi-chlorians.

One problem with the idea that midi-chlorians exist in all complex life, though, arises because we are talking about life on many different planets. As we discussed in Chapter 2, life on other planets is likely to be radically different from life on our own. It's unlikely alien life will even use the same DNA molecules that terrestrial life does to carry its genetic code. So how could the same bacterial ancestor to the midi-chlorians arise on every planet? Remember Dr. Cohen's statement, that "Finding another planet with our kind of dinosaurs or people is more unlikely than finding a remote Pacific island on which the natives speak perfect German." Alien life would have gone through an entirely different evolutionary procedure, suffering different accidents of history. Would alien life have the same type of mitochondria that terrestrial life has? Dr. Gray says, "I certainly wouldn't imagine that they could be the same."

We're in a slightly better position than Dr. Cohen's statement might suggest, since bacteria are very primitive life forms, and so develop fairly early in the evolutionary process. We might suppose that similar life forms would develop on other planets. In fact, scientists feel that the life we'll find on other planets will most likely be simple bacterialike life. Yet this alien bacterial life may be structured very differently from terrestrial bacteria.

If these midi-chlorians are supposed to be identical in all life, a much more plausible scenario than having them develop independently on many planets would be that the bacterial precursor of the midi-chlorians developed on one planet, and then spread to other planets as their hosts traveled to those planets. That would mean life on other planets would be limited to primitive one-celled organisms until the ancestor of the midi-chlorians reached them. And once this bacterial ancestor did arrive, it would most likely be incompatible with the indigenous life, and so would have to start its own evolutionary process on each planet. While this theory is possible, I don't find it very convincing. By the time life evolved to a point where an intelligent species could travel to other planets, my guess is that the midi-chlorians would have lost their ability to survive outside a host, just as our mitochondria have. The bacterial precursor to the midi-chlorians would likely be long extinct, and so multicellular alien life would be impossible.

Instead, let's consider another possibility. Let's imagine that multicellular alien life requires a specialized organelle for producing energy, just as multicellular terrestrial life does. That seems a reasonable possibility. If that is true, then all multicellular life on all planets would have something *like* mitochondria. But to say that exactly *identical* mitochondria would develop independently on multiple planets is extremely implausible. Much more likely, other planets would develop their own energy-producing organelles. And only those planets on which such organelles developed would have complex life. Similarly, if multicellular life requires Force-accessing organelles, then most likely other planets will produce their own version of these. Perhaps what the Jedi call midi-chlorians are actually many different organelles from many different planets all serving the same purpose.

In this way, we can imagine a universe in which all complex life forms have midi-chlorians incorporated into each cell. Could these organelles do what the midi-chlorians are said to do?

"I'VE GOT A BAD FEELING ABOUT THIS"

The midi-chlorians are said to communicate with the Force, and if a person with a high concentration of midi-chlorians is trained, he can "hear" the midi-chlorians telling him the will of the Force. Thus a Jedi might "have a bad feeling" about something or "sense a great disturbance" in the Force.

In these cases, the midi-chlorians seem to be serving almost as an additional sensory organ, putting the host in touch with a force in his environment. Just as we might "sense a great disturbance" if we felt the rumblings of an earthquake, or saw the funnel cloud of a tornado, a Jedi might sense a disturbance if his midi-chlorians somehow revealed a gathering of destructive energy.

We discussed in Chapter 2 how animals sense different things than we do. Dolphins and bats use sonar to locate objects and determine their shapes. Many animals respond to specific odors, which may drive them to mate or to flee. Other animals have even more unusual sensory abilities. Some moles, fish, sharks, and eels have special sensory organs in their heads that can detect changes in the electrical field in the water around them. Some birds, insects, sea turtles, and newts are able to sense the Earth's magnetic field, and they use this to orient themselves and navigate. These animals are able to access information that we cannot. They can detect a force field that is undetectable by us. Sound familiar?

In Chapter 2, I assumed that *Star Wars* humans were the same biologically as terrestrial humans. Now we know that there is at least one difference. *Star Wars* humans have midi-chlorians, without which they could not live. If that is the only difference between us, could these midi-chlorians give their hosts these special abilities? Could mitochondrialike organelles provide an additional sensory capacity?

Let's look first at animals that can detect the Earth's magnetic field. Different species seem to have different ways of sensing this magnetic field. Scientists are just beginning to understand these. Many organisms, including bacteria and sea turtles, have crystals of magnetite, a magnetic mineral, in some of their cells, and it's theorized that these crystals may act as tiny compasses within the animals, allowing the animals to sense the magnetic field. In some cases these magnetic materials seem to collect in a magnetic sensory organ, forming a chain that makes a large compass. In other animals, such as birds and newts, scientists believe photoreceptor molecules in the eyes may be sensitive to external magnetic fields. While they are not magnetic in themselves, they may be paramagnetic, responding to magnetic fields. As the molecules in the eyes respond to magnetic forces, the animals may actually "see" the magnetic field.

Although we notice no distinctive sense organs on *Star Wars* humans, some of them seem to be accessing additional information about their environment the same as these animals are. We can theorize that just as magnetite crystals within some bacteria may allow them to sense the magnetic field, so some special crystal or molecule within the midi-chlorians may allow Jedi to sense the Force. If the midi-chlorians can somehow detect the field generated by the Force, then people with a high enough midi-chlorian concentration, who have learned to "listen" to their midi-chlorians, might detect changes in the Force around them: disturbances in the Force.

Dr. Gray draws an analogy between the midi-chlorians and chloroplasts in plant cells. Chloroplasts, like mitochondria, are organelles within cells that produce ATP. In plants, this process is called photosynthesis. Scientists believe that chloroplasts evolved from a bacterial ancestor much like mitochondria. The chlorophyll molecules within chloroplasts absorb the energy of the sun, and it is then converted into chemical energy. Dr. Gray says, "One might imagine some kind of an energy force like sunlight which could be released by one individual and captured by another, in the same way that chloroplasts capture sunlight." The midi-chlorians, then, may have receptors that absorb the Force and may even use its energy for some purpose, such as allowing a Jedi to jump really high in the air and do a lot of cool somersaults.

It is said, though, that life creates the Force. How might the midi-chlorians not just sense or absorb the Force, but also contribute to it? To understand how this might occur, let's look at animals that detect electric fields, rather than magnetic ones. Some electro-sensitive animals, fish and eels included, sense electric fields by releasing a rapid series of electric pulses—120 to 300 per second—creating an electric field around their bodies. These are not the high-powered shocks that eels deliver to paralyze prey. They are much lower-powered pulses. Why do they do this?

Well, in water, all plants and animals produce a weak, constant electrical field. Most species are unaware of these fields. But electro-sensitive fish can be up to one million times more sensitive to electrical fields than regular fish. When plants or animals enter the field surrounding an electro-sensitive fish, they cause a slight deformation in the field, and the fish detects it. Such a fish can sense the presence of other fish in the dark or in turbulent water, or even find organisms hidden in the sand at the bottom.

If a person not only detects changes in the Force but also generates a Force field around his body, as these fish do, then he will be contributing to the Force, as all life is said to do. He might then sense the presence of others as a deformation of the field generated by his own body, and his presence may be sensed by other Jedi, just as Darth Vader senses the presence of Obi-Wan in *A New Hope*.

Again, though, this assumes that these *Star Wars* humans, who seem like us in every other way, have Force-sensing organs and Force-emitting organs that we do not have (or at least that we don't know that we have). These electro-sensitive fish have well-developed systems that allow them to release electric pulses and detect changes in the electric field, and these systems are anatomically apparent.

We're postulating an additional sensory system, though one is not apparent in any *Star Wars* humans. They would need Force emitters and Force receptors, and those receptors would need to be connected through sensory nerves to the brain, a brain specially designed to analyze these Force impulses. Dr. Gray agrees that the cells would need to be connected through some sort of network. "It's possible there could be some kind of chemical transmitter or communicator that would go from cell to cell and link them all together. Presumably you would need a neural network at the end." Scientists are finding such sensory systems in the animals discussed above, involving electroreceptors and neural pathways conveying the information gathered by them. If we throw away the idea that *Star Wars* humans are biologically similar to us, we can get away with theorizing something like this, but it seems kind of odd that these *Star Wars* humans would look just like us except for several Force-related abilities. It also removes a fair amount of the mysticism from the Force. If such sensory organs existed, scientists and Jedi in the *Star Wars* universe would know of them, and the existence of the Force would likely be known and understood, not a matter of personal belief.

It seems more likely, based on the mysterious nature of the Force, that Force receptors and emitters work in a much more unusual and subtle way, a way that scientists in the *Star Wars* universe haven't yet discovered. If these receptors and emitters, rather than concentrating in specific organs, could work within each cell, like the chloroplasts, and convey their information to the brain through some chemical messengers that might be difficult to detect, this would allow *Star Wars* humans to remain more like us. Information about the Force, then, would be conveyed along unusual pathways, different from the paths normally used to convey sensory information, which could also help explain why people need to be trained to hear the Force.

Even if we accept that midi-chlorians might have these sensory abilities, the problems that we discussed in Chapter 5 still remain. Information about the Force is detected over huge distances, and even detected before an event occurs, violating basic laws of physics. An electro-sensitive fish, on the other hand, can detect changes in the electrical field only over a distance of about two fish-lengths.

One last intriguing tidbit. If you expose a fish to a moderately strong electrical field, it will suffer from unintentional muscle contractions that force it to swim in a certain direction. We might draw a parallel between this and that "old Jedi mind trick," in which a strong dose of the Force compels a person to act in a certain way.

"THE FORCE IS STRONG IN MY FAMILY"

Another characteristic of the midi-chlorians is that the concentration of them in the body seems to be inherited. Luke says that the Force runs strong in his family. Qui-Gon asks Shmi about Anakin's father, presumably to learn what strength of midi-chlorians Anakin may have inherited. Qui-Gon runs no test on Shmi, which suggests that one's father is particularly important in inheriting a strong connection to the Force.

What's interesting is that exactly the opposite is true of mitochondria. Human eggs and sperm both contain mitochondria, but once fertilization occurs, the fertilized egg destroys all the mitochondria from the sperm. Only those from the egg survive, so the mitochondria you have came exclusively from your mother, and the DNA within that mitochondria is solely your mother's DNA, with no contribution from your father. This is true in most animals. If midi-chlorians developed the same way, then your father would have no effect on the strength of your midi-chlorians. Only your mother would matter.

Perhaps midi-chlorians work the opposite way, with all those of the mother being destroyed upon fertilization, so only those of the father are inherited. This is true of mitochondria in a few species.

Could one person inherit "stronger" midi-chlorians than another person? Certainly, just as someone can inherit stronger mitochondria than another person. The genes in your mitochondria might be healthy and intact, or they might be mutated and damaged. Over one hundred human diseases are caused at least in part by defects in mitochondrial genes. Mitochondrial genes are prone to mutate ten to twenty times faster than the genes carried in the cell nucleus. That's because the job of the mitochondria, energy production, is a dangerous one, and can damage the mitochondrial DNA. As this DNA sustains damage, the energy available to the cell declines. If this occurs in tissues that require a lot of energy, the tissue may die. Or the tissue may survive but lose some of its ability. This mitochondrial decay is believed to contribute to aging. So depending on how healthy your mother's mitochondrial DNA is and how much damage it has sustained, you can inherit mitochondria that may be strong or weak. In the same way, you might inherit midi-chlorians that are strong or weak.

Another important element controlling one's strength with the Force is the concentration of midi-chlorians. We're told Anakin's midi-chlorian count is 20,000, higher than anyone else known. We might assume that means he has 20,000 midi-chlorians per cell. Why might a person have more midi-chlorians than another?

An average cell contains hundreds of mitochondria, though the number varies widely depending on how much energy the specific type of cell needs. A sperm cell contains only fifty to one hundred mitochondria—just enough to give it the energy it needs to reach the egg. An ovum or egg contains about 100,000 mitochondria—enough to begin to build a new human being. Yet from one person to another, the concentration of mitochondria is fairly uniform. Your mitochondria might be stronger or weaker, but you'll have about the same number in one of your liver cells as I have in one of my liver cells.

The number of mitochondria in a cell is regulated by the DNA within the nucleus. If excess numbers were somehow produced, the nucleus would stop the production of more mitochondria until the numbers went back to their normal levels. The midi-chlorians seem to operate differently, since apparently people can have widely differing midi-chlorian counts. Thus we might imagine that one person's nuclear DNA may have different directions for its midi-chlorians than another person's. Anakin's DNA may tell his midi-chlorians to multiply until there are 20,000 midi-chlorians per cell. Boba Fett's DNA, on the other hand, may tell his midi-chlorians to stop dividing when there are fifty midi-chlorians per cell.

Thus one's strength in the Force would be determined by a

combination of the DNA in the midi-chlorians, which controls the quality or strength of the midi-chlorians, and the DNA in the nucleus, which controls the quantity. Many fans have speculated that George Lucas introduced the midi-chlorians to help set up a plot twist to come in one of the future movies. Let's examine the implications of having midi-chlorians as carriers of the Force.

If we wanted to artificially increase someone's strength in the Force, it would not be enough to inject him with a load of midi-chlorians. Aside from the fact that the body would reject these foreign invaders, the concentration of midi-chlorians would quickly decline as the nucleus of each cell stopped their reproduction until the numbers fell to normal levels.

If we wanted to create a person strong in the Force, we could not simply clone a Jedi, at least not a male one. In cloning, the nucleus is removed from an ovum, and in its place is put the nucleus from a regular cell of the person to be cloned. If we wanted to clone Anakin, say, we would take an ovum from a donor, remove its nucleus, and insert the nucleus from one of Anakin's cells. Such an ovum could then be stimulated to grow into a clone of Anakin. The nuclear DNA of this clone would be identical to that of Anakin. Yet the mitochondrial DNA—and likely the midi-chlorian DNA as well—would be that of the donor, not Anakin. To create a full clone of Anakin, with the same quality of midi-chlorians that he has, we would need to transfer not only the nucleus of one of his cells but also the mitochondria and midi-chlorians. This could be done— the mitochondria transfer has actually already been done in mice—but it would be a more complex process than normal cloning. With such a process, we could potentially create new Jedi with the same strength in the Force as Anakin. Which could make the Clone Wars really interesting.

ARE YOU SURE YOU DIDN'T JUST GET DRUNK ONE NIGHT?

The most controversial aspect of the midi-chlorians is their alleged role in the conception of Anakin. According to Anakin's mother, Shmi, Anakin had no father. While my first tendency was to disbelieve her, and that is still my preferred option, I don't believe that is George Lucas's intention. He presents Shmi as an honest and intelligent woman, one we should probably believe. While many fans have presented other alternatives—Shmi was drugged and forgot she had sex, she was artificially inseminated without her knowledge, she was abducted by aliens, and more—again, these seem to go against what George Lucas wants us to believe. Shmi says she can't explain what happened. Can we? Is it at all possible that a child could be born without a father?

Such a process does occur in some animals. It's called parthenogenesis, which in Greek means virgin birth. In parthenogenesis, an ovum or egg develops into a complete offspring without the fertilization of a sperm. A variety of animals can reproduce this way, including insects, reptiles, and domesticated turkeys. Some species that reproduce through parthenogenesis are all females, which means they have no choice but to reproduce without a male. Other species leave their options open, reproducing both by parthenogenesis and by more conventional sexual means.

Turkeys are one of the latter. They reproduce through parthenogenesis when there is some problem with normal sexual reproduction, for example, when the male turkey's sperm count is low, or when the turkeys are infected with viruses. When turkeys reproduce this way, the offspring produced are all male, and only 20 percent of these males are fertile.

In species where parthenogenesis does not naturally occur, it can be artificially stimulated. This has been done in a wide variety of animals. The unfertilized egg has to be tricked into thinking that it has been fertilized, so that it will begin to divide and grow. In 1900, Jacques Loeb pricked unfertilized frog eggs with a needle and found that some of the eggs developed as if they had been fertilized. Rabbit eggs have been stimulated with temperature and chemical changes. Yet most of these artificially stimulated eggs develop abnormally. And all such mammal embryos stop developing before becoming fully formed. No cases of human parthenogenesis, natural or artificially stimulated, have ever been scientifically documented. Yet could this ever happen?

Just two years ago, researchers discovered five possible cases of parthenogenesis in snakes, which was never before suspected. At first scientists were skeptical, because female snakes have been known to store sperm for years before using it to reproduce. But female snakes were found giving birth to young when some of them had never been exposed to male snakes. Most of the offspring died as fetuses, but several were born alive and have survived. DNA tests revealed that the DNA of the offspring came solely from the mother. Scientists now believe that snakes may reproduce this way when mates are unavailable. So if a female can't find a male to reproduce with her, she can do it by herself.

Is it possible that Shmi, if kept away from human males for some years, left only with Watto, Jabba, Bib Fortuna, and the gang, would reproduce on her own? Scientists now believe a special trait of mammalian DNA prevents parthenogenesis from occurring in all mammals, including humans. The DNA of a sperm and the DNA of an egg seem to be imprinted differently, and an embryo will only develop if both imprintings are present.

Even so, parthenogenesis would be the simplest explanation for Anakin's birth, except for one thing. Anakin is a male. If parthenogenesis did occur in a human, the result should be a female. In parthenogenesis, the child is essentially the product of a mating between an ovum and itself. Since the ovum has half of the genetic material of the mother—which is normally combined with half of the genetic material of the father—it can include only traits that are seen in the mother, with the exception of any genetic mutations. The sex of a human, as with all mammals, is determined by X and Y chromosomes. A chromosome is just a long string of DNA that is wound into a tight bundle. A female has two X chromosomes. A male has one X and one Y. Since the child could have only the DNA of the mother, it could not have a Y chromosome, and so would have to be female.

Which leaves us with the big problem of Anakin. Of course, there is another famous male figure who was allegedly a product of virgin birth.

To come up with any possibility, we have to get a little creative—actually, a lot creative. There are a couple of different possibilities we could explore, all of which are extremely unlikely. After all, if it were likely, we'd see a lot more women giving birth without the participation of men. I'm going to discuss the possibility that seems to be the most likely, with the understanding that even the "most likely" is still very, very unlikely.

A condition exists in most animals and humans called mosaicism. Let's go back to the moment of your conception, when your life began as a single fertilized egg. That egg contained the DNA that held the plan for your body. As that egg divided and divided into many cells, the DNA was copied so that each cell received an identical set. At least that's how it's supposed to work. In practice, your DNA is not copied perfectly every time. If an error is made, creating a cell with a bit of faulty DNA, then as that faulty cell divides and divides, all its future generations, all the cells in its *cell line,* will carry that same fault. Now you'll have two types of cells in your body. One group will carry the original DNA plan for your body; the other will carry the altered, faulty DNA plan. Thus you are now mosaic; all your cells no longer carry the same DNA.

Only in the past few years have scientists begun to understand the implications of such mosaicism. It's believed now that all people are mosaic, since errors or mutations in DNA occur on average in one in every million cell divisions, and we have ten trillion cells in our bodies. The health consequences of the mosaicism depend on what type of fault the DNA carries, how many cells arise in that cell line, and the locations of these cells.

For a few people, genetic mosaicism can begin at the very moment of conception and can involve much more radical genetic differences. An ovum may, through malformation, have two nuclei. These two nuclei can be fertilized by two different sperm. The two fertilized nuclei then both grow and divide, contributing cells to the developing embryo, though these cells are genetically different. Some cells in the body will have one DNA plan for the body; other cells will carry a different DNA plan. It's as if some of your cells were combined with some of your sibling's cells into one body. Such cases have been documented in animals and in humans. Dr. Alan Beer, professor of obstetrics and gynecology, microbiology and immunology at the Finch University of Health Sciences/Chicago Medical School says, "No one would dispute this possibility." Obviously this can potentially cause serious problems in the fetus. If the health problems are not serious, more mild signs of mosaicism may be found. A baby may have one leg longer than the other, or one ear higher than the other. But if most of the cells that form the baby share the same DNA, and only a small number have the secondary DNA, the baby can appear normal, and the mosaicism may never even be discovered.

Let's say now that Shmi has such mosaicism. Most of the cells in her body share the same DNA, and that DNA is perfectly normal. Yet a few cells in her body have different DNA. These different cells are in her ovaries, and make up some of her ova. Again, this type of mosaicism has been documented in humans. In many cases, it goes undetected until two parents who seem free of a particular dominant genetic disease give birth to a child who has the disease. The genetic coding for the disease existed hidden in the parent's ovum or sperm.

But mosaicism isn't quite enough to explain the birth of a child without a father. For that, we not only need two groups of cells in Shmi's body, but the second group, the cells that make up only a tiny proportion of Shmi's body, including some of her ova, must have DNA that is not only different, but abnormal. Hey, I said we had to get creative! Stick with me on this.

Normally, ova and sperm each have twenty-three chromosomes, half the number of chromosomes of a normal cell. Then when an ovum and sperm meet and create a fertilized egg, the egg has forty-six chromosomes, the total number that should appear in each normal cell in the body.

Once in a great while, two sperm can fertilize the same ovum. The resulting fertilized egg will have sixty-nine chromosomes. This is called triploidy. Normally, the first sperm to reach the ovum sets off a reaction that strengthens the ovum's cell membrane, preventing any further sperm from entering. But, as Dr. Beer explains, "If two sperm reach the finish line at the same time, both can go in." Two percent of all conceptions are triploid. Of those triploid conceptions, only one in 1,000 is live-born. And of those, the average lifespan is only twenty hours. Such a baby has three sex chromosomes rather than two.

A fetus with sixty-nine chromosomes would not normally survive and grow. Yet, if only a few of the fetus's cells had sixty-nine chromosomes because the fetus had genetic mosaicism, then it could live and grow into a baby, with most of its cells normal and healthy. Yet inside would lurk these few abnormal cells. This may be what Shmi is like.

So let's review what might have happened at Shmi's conception. Shmi's mother may have produced an ovum with two nuclei. Of those two nuclei, one was fertilized by a single sperm; the other was fertilized by two sperm. The first fertilized nucleus had forty-six chromosomes, including sex chromosomes XX. This nucleus developed normally, making up the majority of cells in Shmi's body. The second nucleus had sixty-nine chromosomes, with sex chromosomes XXY, and made up only a small portion of the cells in Shmi's body, including some of her ova.

We then have, within the same ovum, two unusual events —the ovum has formed with two nuclei, and one of those nuclei is fertilized by two sperm. Is that reasonable? Dr. Beer believes that the two events may actually be linked, that it may be more likely for triploidy to occur in an ovum that has two nuclei. "It's quite possible that these dual-nuclei eggs are more prone to polyspermia. Their cell membranes may have a greater susceptibility for sperm to enter."

If this is what happened at Shmi's conception, then a very interesting thing could occur in the adult Shmi. Ova are created when a cell splits through a process called meiosis, a process in which a cell with the normal number of forty-six chromosomes splits to produce two ova, each with only twenty-three chromosomes.

Yet if one of Shmi's cells had sixty-nine chromosomes, it might very well split so that one of the resulting ova had the correct number of twenty-three, while the other had the full number that a fertilized egg would have, forty-six. The ovum with forty-six chromosomes could have the XY pair of sex chromosomes necessary to create a male. This ovum would then have all the ingredients it needs to create Anakin.

Even so, the ovum would know that it was not a fertilized egg. It would need to be tricked into thinking it was, just as in parthenogenesis. Could this perhaps be the role of the midi-chlorians: to stimulate this cell so that it would believe it had been fertilized and would begin to divide and grow into an embryo? Dr. Beer explains that "some kind of stimulus is required: electric shock, osmotic shock where we change the nature of the cell membrane, or even pricking the cell surface with a surgical probe."

Qui-Gon theorizes that the midi-chlorians played a role in the conception of Anakin. A midi-chlorian, or a mitochondrion, could not play the role of a sperm, providing half of Anakin's nuclear DNA, because it would not have the same DNA as the nucleus. Yet if all the nuclear DNA is provided by Shmi's abnormal cells, perhaps the midi-chlorians could somehow alter conditions within the cell to trick it into believing it had been fertilized. Could the midi-chlorians potentially provide the required stimulation? Dr. Beer says yes. We could even postulate that the midi-chlorians played some role in creating Shmi's genetic mosaicism in the first place.

Why would the midi-chlorians go to all this trouble? If the midi-chlorians are inherited through the mother, as mitochondria are, then it could be a lot of trouble for nothing. Any man who might be Anakin's father would have no effect on the quality of Anakin's midi-chlorians. Anakin's midi-chlorians would have the same DNA and the same quality as Shmi's. Which might make us wonder why Shmi isn't the chosen one. If the midi-chlorians are instead inherited from the father, then they could have a reason for making Shmi reproduce without a father, who might provide weak midi-chlorians. Yet in that case, Anakin's midi-chlorians would again have the same quality as Shmi's.

The likely answer comes from our earlier theory. While the DNA within the midi-chlorians may control their quality, the DNA in the nucleus may control their quantity. In that case, it is easy to understand how Shmi's midi-chlorian count might be normal, while Anakin's is not. While Shmi and Anakin both, in essence, have the same parents—Shmi's parents—they carry different nuclear DNA, just as you and your siblings do. I am left-handed; my sister is right-handed. Similarly, I might have the gene for a high midi-chlorian concentration, while my sister does not. And so while Shmi's midi-chlorians may be high in quality, they may be low in quantity.

We have one final element to worry about in the birth of Anakin. Dr. Beer points out that we don't want this ovum to begin division and form baby Anakin within the ovary or fallopian tube. "You'd get an ovarian pregnancy or fallopian tube pregnancy which would kill the mother. You want this cell to be shocked into division in the uterus." So the midi-chlorians would have to be smart enough to provide their stimulus at the right time.

Thus a child might be born without a father, and might carry traits different from his mother. This event requires three rare occurrences: an ovum with two nuclei; fertilization of one of those nuclei by two sperm; and stimulation of an unfertilized ovum so that it begins to divide and form an embryo. Is it possible? Dr. Beer says, "Based on everything you've told me, this is biologically possible, but it would be a very unusual and unique event. In my estimation it would require an external event—magnetic field, power source, divine intervention, something or other—to stimulate the ovum to begin division." If the midi-chlorians can provide that stimulation, then we have an unlikely but scientifically possible explanation for Anakin's birth.

George Lucas's introduction of the midi-chlorians as the microscopic life forms that allow all species to connect to and contribute to the Force presents some intriguing new scientific avenues to explore. And perhaps episodes two and three of the Star Wars saga will reveal even more. Ultimately, the Force remains a great challenge to understand scientifically. But then all the forces in our universe have posed great challenges to scientists. Isn't that what makes trying to understand them so fascinating?

RECOMMENDED READING

Bohm, David. *Wholeness and the Implicate Order*. London: Routledge & Kegan Paul, 1980.

Cyberlife Technology. *http://www.cyberlife.co.uk* (29 Nov. 1998). The home page of Cyberlife, the company that produces Steve Grand's game "Creatures."

Damasio, Antonio R. *Descartes' Error: Emotion, Reason, and the Human Brain*. New York: Putnam, 1994.

Druckman, Daniel, and John A. Swets, eds. *Enhancing Human Performance: Issues, Theories, and Techniques*. Washington, D.C.: National Academy Press, 1988.

George, Uwe. *In the Deserts of This Earth*. New York: Harcourt Brace, 1977.

Hyman, Ray. *The Elusive Quarry*. Buffalo, N.Y.: Prometheus Books, 1989.

Jakosky, Bruce M. *The Search for Life on Other Planets*. Cambridge, England: Cambridge University Press, 1998.

James Randi Educational Foundation. *http://www.randi.org* (12 Nov. 1998). The website of James Randi (The Amazing Randi) and the James Randi Educational Foundation. You will find here the rules regarding the $1.4 million-dollar reward for "the performance of any paranormal, occult, or supernatural event, under proper observing conditions."

Journal of Scientific Exploration. *http://www.jse.com/JSEHomt.html* (8 Feb. 1999). If you like to read about strange experiments into the paranormal, this is the magazine for you.

Kaku, Michio. *Hyperspace*. New York: Oxford University Press, 1994.

Marks, David, and Richard Kammann. *The Psychology of the Psychic*. Buffalo, N.Y.: Prometheus Books, 1980.

Picard, Rosalind. *Affective Computing*. Cambridge, Mass.: MIT Press, 1997.

Pickover, Clifford. *The Science of Aliens*. New York: Basic Books, 1998.

Randi, James. *Flim-Flam! Psychics, ESP, Unicorns, and Other Delusions*. Buffalo, N.Y.: Prometheus Books, 1988.

Stenger, Victor J. *Physics and Psychics: The Search for a World Beyond the Senses*. Buffalo, N.Y.: Prometheus Books, 1990.

Stenger, Victor J. *The Unconscious Quantum*. Buffalo, N.Y.: Prometheus Books, 1995.

Yapko, Michael D. *Essentials of Hypnosis*. New York: Brunner/Mazel, 1995.

Zuckerman, Ben, and Michael H. Hart, eds. *Extraterrestrials: Where Are They?* 2d ed. New York: Cambridge University Press, 1995.

INDEX

ABOUT THE AUTHOR

Jeanne Cavelos is a writer, scientist, editor, and teacher. She began her professional life as an astrophysicist and mathematician, teaching astronomy at Michigan State University and Cornell University and working in the Astronaut Training Division at NASA's Johnson Space Center.

Her love of science fiction sent her into a career in publishing. She became a senior editor at Dell Publishing, where she ran the science fiction/fantasy program and created the Abyss horror line, for which she won the World Fantasy Award. In her eight years in New York publishing, she edited numerous award-winning and best-selling authors.

A few years ago, Jeanne left New York to pursue her own writing career. She is the author of *The Science of* The X-Files and of the *Babylon 5* novel *The Shadow Within*.

Jeanne is also the director of Odyssey, an annual six-week summer workshop for writers of science fiction, fantasy, and horror held at New Hampshire College. Jeanne also teaches writing and literature at Saint Anselm College.

You can visit her website, which contains additional discussions of scientific issues raised in *Star Wars*, at *www.sff.net/people/jcavelos*. She welcomes your comments at *jcavelos@empire.net*. You might even get her to tell you about some of the bizarre and embarrassing things she's done as a dedicated *Star Wars* fan.

CPSIA information can be obtained at www.ICGtesting.com

224875LV00002B/24/P

9 780312 263874